Beaut Ute

Lothian
BOOKS

Allan M. Nixon

Cover: Special thanks to all those ute owners who turned up at the Guildford Store, in central Victoria, early one morning for the cover shoot. Some travelled many miles to get there, and more turned up than we expected. The word had got around. For their keenness, cooperation and patience I offer my sincere thanks. Some of those utes included:

front left to right: 1953 Morris Minor (silver grey); WB Holden (cream); FJ Holden (silver grey); XR6 Falcon (red); FB Holden (grey)

rear left to right: 1954 Ford F100 (blue-green); HZ Holden (black); Toyota LandCruiser (yellow); Toyota HiLux (blue).

I would also like to thank John and Jacinta, owners of the Guildford Store and the Lodden B & B; and the two 'old codgers' who are local identities — Trev Taylor (in braces) and Bill Mein (in hat). Sadly Bill's dog Podge, sitting on the back of the WB Holden has since died. The other dogs are Rusty (back right), and the 'star' in foreground is Ky, the wonder dog.

Dedicated to the memory of my uncle
KEVIN JOHN GRINHAM
'The Bush Philosopher'
1935–1998
A great mate always

Thomas C. Lothian Pty Ltd
11 Munro Street, Port Melbourne, Victoria 3207
Copyright © Allan M. Nixon 1998
First published 1998. Reprinted 1998, 1999.

National Library of Australia
Cataloguing-in-Publication data:

Nixon, Allan M., 1951–.
   Beaut utes.

   ISBN 0 85091 944 4.

   1. Motor vehicles–Australia–History.
   2–Australia–Anecdotes. 3. Trucks I. Title.

629.223

Design by Geoff Hocking & Allan Nixon
Printed in Australia by Hyde Park Press

1934 Ford

1948 Holden

Contents

A Boyhood Memory – 4
People with Utes are a bit odd – 5
An Aussie Invention – 8
Amanda & Ben – 10
A Swagman & 3 Utes – 12
Austins — A Family Affair – 15
'Aussie Earl', The – 16
Big Black Mother – 20
Blacksmith – 22
'Bluey' – 24
Bronte – 26
'Burkie' — Bogged, Burnt & Shot – 28
Chookologist – 31
Chris, Rusty & The Rig – 32
Claire – 35
'Col' – 37
'Concrete Charlie' – 38
Darwin & Beyond – 41
Emu Eaters B & S;
   Beer, Bundy & A Mug on a Rope – 43
15 dead Emus, 90 howling Dingoes,
   3 barking Dogs ... & a Holden Rodeo – 48
Falcon — Mark Moretta – 51
'Feral' – 53
Dogs, dogs, dogs... – 55
'FJ' – 56
Frank the Rabbiter – 58
'Fuzz' – 59
Grace & the Honeymoon Ute – 61
In Search of... – 64

Gravedigger – 65
Graveyards – 66
Great Divide Ute – 68
Grunter Hunter – 71
Gulgong Jimmy – 72
Half Million Mile Mailman – 74
Hard-Rock Miner
   & the Outback Brumby – 75
Harleys & a Holden – 77
'JB' & the Red Hillbilly – 78
Jock's Mercedes – 80
Jolly Green Giant – 84
Karen's V8 Vauxhall Velox – 85
Mallee Boy — John Williamson – 89
Merv – 95
Mini Magic – 96
Morrie Mania – 98
Mudbricker – 101
Natalie – 102
Outback Cop – 104
'Paddock Basher' – 105
'Pass The Test' – 106
Peter Lamb's 1962 XL – 109
Plumber 'Baz' – 110
'Pop' – 111
'Queen of Speed'
   — Rachelle Splatt – 112
Reo Speedwagon – 116
Republican Ute – 117
Roadwarriors – 118

RR 'Silver Ghost' – 120
Ron's Chev – 122
Runt of The Litter – 124
Rust to Restored – 125
'Sarge' & Vanessa – 126
Shirl's 2-Tone Mark II – 128
Simply... Peaches, Pears, Prunes
   & Wombat Hubcaps – 130
Steel Stouts – 131
'Stewie' & The Apprentice Ute – 133
Signman – 134
Studebaker – 136
Stylemaster – 138
Swamp Dodge – 140
The Bloody Mongrel – 141
The Flying Sculptress – 144
The Gums – 145
T-Model Jackaroos – 146
Tony & Deidre's '54 Ford F100 – 148
'Top Gun' – 149
Vanguard – 150
Versa-U-Tility – 152
Vikki's 2nd Ute – 153
Wandering Sparkie – 154
Wayne's 29er – 155
Working Man's Rolls Royce – 156
WBs – 158
Young Guns – 159
Acknowledgements – 160

Before I was old enough to go to school, I grew up in the back of a Ford '10'. Much of my daily life was spent in the Ford seated on a wooden grocery box surrounded by full boxes of groceries. I had a threepenny bag of mixed lollies from our grocery shop to keep me amused, and probably to take my thoughts off the bone-jarring and mind-numbing ride.

This was my 'home', as my mother drove around with me and young 'Twinks' the grocery boy, who would run into households that had ordered from our family grocery and hardware store in our old gold town in Central Victoria.

In later years our shop owned a big cream/light brown 1946 Chevrolet ute. I'd spend hours 'driving' the ute whilst parked in the shed, changing gears, even though I was still too small to reach the clutch, brake or accelerator. Kids have great imaginations!

As I grew older, after school I would go on delivery runs seated beside Twinks, often carrying boxes of groceries and unloading them onto some woman's kitchen table, scooping up the money and returning to the ute that was still running.

Deliveries often meant bread and jam or soft drink as a reward. I remember delivering to unusual places such as the old swagman's hut on the edge of town, or to the man who had two big pickle jars of gold nuggets, or to the woman with the wonderful kerosene lamps in every room.

We went everywhere in the Chev. I loved that ute. Life in the 1950s was simple — no locked doors, trust in your neighbours. Dogs sometimes bluffed me though, and Twinks would come to the rescue. Great bloke Twinks.

As years passed I would own my own utes. A great one-owner 1964 Falcon that I paid $500 for, a black vintage 1927 Chev, XF and XG Falcons. The list is growing, and many more are on the way, of course! An FJ Holden is one of the priorities.

My parents remember the old Ford '10' as a real bone shaker, but nothing could stand in the way of a young boy and his bag of lollies!

'People with Utes are a bit odd — I don't usually associate with them'.

This was an unexpected response from an Aussie Italian mate of mine. Everyone has different ideas about utes. I can probably safely say I've looked at more utes than anyone else in Australia having driven many thousands of kilometres all over the place looking at all sorts of utes and their owners. That's all I see on the road now; I'm oblivious to anything without a rear tray.

Tony was right in one respect about ute owners, I've found. They are different when it comes to their vehicles. Ute owners don't just drive a ute — they are generally passionate about their utes. One bloke summed it up: **'It's a heap of shit, but I still love it'**, a sentiment echoed by others.

Some people give their utes names like 'the old girl', 'Jolly Green Giant', 'Wobbles', 'Desmond Datsun' or 'Smokey'. They talk to them, enjoy driving them, make love in them or, on the other end of the scale, treat them roughly, abuse them, take them for granted, overload them with every imaginable thing and, as one man said, **'work the arse off it'**.

Most utes are 'big boys' toys' but the girls love 'em too, especially if the ute has 'the look'. It is a status symbol for country girls to own a ute. The 'special' ute attracts admirers who drool over its qualities.

Most ute owners wouldn't part with their utes for anything; some say they'd never sell it, no matter what. **'Geez mate, take the house, the missus and the kids, but leave me dog and ute!'**

Utes come in all shapes and sizes. Arguably an Aussie invention, they are under threat of being swamped with models from around the world. The current new 'invasion' is the big bloody 'yank tanks' known as pickups in 'goddamn' land.

But there will always be a resistance movement here in Australia, as long as the kookaburra sings, the magpie carols and Skippy reigns. And as long as the throbbing bark of a V8 Holden or Falcon rings out across the land.

The Good, Bad & Ugly

Utes. We accept them as part of our everyday life. We talk about them and swap parts for them. We put stickers on them for fun, and signs that advertise. We flog them and wash them, and kill them, leaving them to lie carelessly in a paddock with rust setting in.

We mould them and revive them and bring back the shine of chrome and paint and restore them. We trash them and sell off the bits to the wrecker or at swap-meet markets. Life without them just would not be the same.

Utes can be just a bomb held together with fencing wire. Stock crates or tarps, canopies of canvas or fibreglass cover the tray. Riding in the back are dogs that bark to protect what's 'mine', empty Foster's beer cans on the floor, Bundy bottles too. High-profile looks with chrome and paint and mags, or just plain rust, it really doesn't matter much so long as your ute has your stamp on it.

Old Codgers

Usually you see them coming, slowly. Slouched over the steering wheel, wearing a lazy 'how it sits is where it stays' kind of hat, old and battered, the old codger, his hat and ute are inseparable. They've all seen better days. Hard days. Hot days, and

wet ones too, mud caked on so thick underneath the mudguards that it holds the rust together.

Sometimes he has 'Mum' sitting beside him as he arrives in town, in the floral frock she made in about 1973, a hat, and a handbag. Always a handbag. Usually though it's a kelpie beside him, sometimes more than one, the young ones on the back tray, barking or running from side to side, slobbering excitedly, full of importance. He'd probably rather have his dogs in the ute than Mum anyway. She makes a trip into town more of an effort. Grocery shopping for this and that, here and there. He'd be just as happy to get to the bank, then down to the co-op for a few things for the farm, then he'd be free to lean on the mudguard outside the Royal Arms and have a good yak about current sheep market prices.

The old lady farmer is there too, with pride and joy in her heart, thinking nothing of being in her Sunday best behind the wheel. She bought the ute brand-new and has driven it for years. She loves her ute!

Young Bucks

Born and bred in the bush. Farmer of the second or third generation stock, or even more. Rough and tough. Casual but sometimes full of importance. Boots and hats always a trademark of the young buck bushie, just like the ute and dog. Hard worker, hard drinker, hard lover. Doesn't go for those city blokes and their shiny revved up machines much. Young bucks prefer real working machines like Holdens, Falcons and Toyotas.

Stickers play a big role. Macho ones like 'Bulldoze a Greenie'. Likes to stick it up the trendies with a parody of their sticker: 'Wankly-thermonuclear Ejecu-lation' instead of 'Oakley thermonuclear protection'.

Country boys love all types of music, from Jimmy Barnes to John Williamson. Mostly they love Country Rock though.

City young bucks. Cruising. Always cruising. Chrome and shining duco and the right sound are key elements of a city ute. The thump, thump, thump, music always loud, heard three blocks away before you see it coming. 'P' plate in the the window. Stickers are usually 'No Fear' or smart-arse stickers like 'Get in, sit down, shut up and hang on' across the full width of the rear window.

Sheilas

Sheilas are a different breed when it comes to their vehicles. Often stickers show attitude — 'Crazy Bitch with PMT' emblazoned in bright red right across the rear window or other 'personalised' words of defiance. They are often hogging the right-hand lane on freeways, stubbornly refusing to move over for anyone!

Bush sheilas will shove a sticker on their vehicle, and will share a swag under the stars. Not so the city girl. City girls like comfort and cocktails or spirits with a bang; the country girl will be just as comfortable with a beer or fluffy duck. Modern sheilas can be extremely confident and good underneath the bonnet, some becoming qualified mechanics, challenging the once traditional male-dominated workplace.

This is a book about Australians. Great ones. Individuals. It is about their love of freedom, their much loved dogs and about their Aussie invention — the ute. It is more than that though. It is ordinary people who go about daily lives in which their utes, and dogs, play a big part. It is about their passions, their work, their mates.

I make no apologies for the fact that this book may sometimes be seen as politically incorrect by some. I hope so, because the story of everyday Australians has to use their language to tell it as it really is. I believe we should, one and all, **'Talk the Slang, Not the Twang.'**

If you speak politically incorrectly with no malice intended, with laughter in your heart and respect for others in your mind, no one should be offended, and should understand that the other is expressing the right to freedom of speech. We are, after all, a country of great diversity, so let us retain that diversity as well as our uniqueness.

This book is full of old codgers, young blokes and sheilas. Their dogs Bluey, Snifter and Wingnut and others all crack a mention. There's reference to screwing, getting pissed, having a good time, hard yakka and the daily grind. It is about the last frontier country, patriotism, and identity. It is about utes.

I hope you get a good laugh, a warm feeling about Australia and Aussies, both new and old, and a greater respect for mates, their dogs and the humble but mighty ute. Space does not allow us to cover every make and model of ute. I hope you like the ones we have included.

Rip into it and enjoy it as much as I did chasing utes and bringing them to your attention. Hopefully, you'll never look at a ute the same way again.

For a future book on utes, write to me with details of your own favourite 'ute-of-all-utes'. Include a photo if possible. Write to my Melbourne office: c/- P.O. Box 46, North Essendon, Vic. 3041. I'd love to hear from you.

May your wheel nuts always be tightened. See ya' in the Spring!

Allan M. Nixon
The Uteman

P.S The Uteman website is under construction. Keep an eye out on the Internet for it in the future.

AN AUSSIE INVENTION

I can hear some of you purist ute owners saying 'that's not a ute' to a couple of those in this book. But when is a ute a lorry or light truck, standard ute, a roadster ute, flat top, stakeside, buckboard, dropside, well type, tabletop, express, cab chassis or coupé utility? In early catalogues and advertisements of the 1930s, the word lorry was often used. Pickup, of course, was and still is an American word; let's hope it is never used in Australia.

In the early days passenger vehicles were cut down, their rear tubs removed by their owners and a custom-built rear tray added. Bastardisation will happen to all makes and models of vehicles as long as there are cars on the road.

Americans will tell you that America's first ute was a 1931 Ford closed-cab pickup, called the Deluxe Pickup. Shipments of these vehicles started going out on 7 May 1931, but a total of only 293 units were built and this pickup regarded a production failure. Apparently it was created because General Electric put in a large order for a fleet sturdy enough to carry refrigerators and other heavy appliances.

America's 'first truly successful coupé utility' became the Ranchero, not unveiled until November 1956 and advertised as a 'car 'n truck rolled into one'.

At Ford Australia in 1934 at Geelong, the first one-piece body 'Coupé Utility' was completed and a couple were shipped to Canada. One story says that Henry Ford showed some Texas cattlemen, who called it the 'Aussie Kangaroo Catcher'. Another article says Henry Ford summoned his men from Texas to inspect the utility, which he jokingly referred to as a 'kangaroo chaser'. The real story is lost to motoring folklore.

Most American pickups that followed in the mid-1930s to late '40s still differed from the genuine Aussie ute in that they had a separate pickup bed and cab.

The Aussie ute designed in 1932 by 22-year-old engineer, Lew Brandt, came off the production line in 1934 and was an instant success, unlike the American pickup. It was the forerunner of millions of similar vehicles manufactured worldwide over the next 50 or more years. Henry Ford wrote acknowledging and thanking Lew Brandt for his design.

There are many different claims about the ute, and I will not devote time to them all here. You will find a good representation of what most would consider a ute, as opposed to a sedan, in *Beaut UTES*. I'll leave the arguments to the 'experts'.

We certainly can claim to have invented **the world's first successfully mass produced one-piece body 'coupé utility'**. It began with a letter of complaint from a farmer's wife in Gippsland, Victoria, who said her husband wanted a vehicle that could carry her to 'church on Sundays and pigs to market on Mondays'.

The rest is history, thanks to brilliant young Australian designer Lew Brandt.

THE NEW FORD COUPE UTILITY 302

FEATURES OF THIS SMART NEW COMMERCIAL VEHICLE
ARE COMFORTABLE PASSENGER ACCOMMODATION
AND LARGE CARRYING SPACE IN REAR COMPARTMENT

AMANDA & BEN

I was going to call them the Jack 'n Jill-a-roos, but Ben no longer chases sheep and cattle, instead he drives a large header, harvesting all sorts of crops such as corn, barley, soy beans, canola and sorghum.

They met at Murrumbidgee Agricultural College at Yanco, New South Wales.

Amanda originally came from Bourke, New South Wales, then Hay. She's worked as a barmaid and in a Pizza Hut but her real ambition was to work outdoors as a jillaroo. Luckily she saw an advert in a newspaper for a rural traineeship. Now she is into her third year working on the same large property. She gave her Nissan Bluebird to her parents and bought a brand-new 1997 Ford Longreach ute, which she calls 'Chucky'. A sign on the front windscreen says 'Little Miss Mischief'.

'I absolutely love it one hundred per cent. A ute has got style. It has class. I've always wanted one. One day in a garage a bloke asked what I drove, and when he saw the ute, he looked at me in amazement and said "I didn't think girls owned utes. Boy, things have changed a bit." I just laughed. Mum and Dad couldn't understand why I wanted a ute, but they didn't try to talk me out

of it. In the end they were both pretty positive. I guess I've always been a bit of a tom-boy. Always.'

Amanda's workdays now are spent in a farm ute with the dogs Butch and Rusty, or on a motorbike or horse. She does everything, from chasing cattle and sheep to crutching, watering crops, sub-clovers and pastures, to fencing, general maintenance, and ploughing or sowing on a tractor.

Amanda's boyfriend, Ben, works on another property. He used to work on a property of 105,000 hectares (260,000 acres) doing stock work, later working with cotton. He now works out in a part of western New South Wales where the dust blows, as the back of my ute can testify.

His ute is a WB Holden and it is his first. He bought it privately about twelve months ago for $6500, and has spent 'about that or

more on it since'. It's got a 253 V8, bullbar, mags, stereo, UHF radio, and it has *worked*. 'I had a prang in it and landed in a drain. I ended up in the passenger seat, but I was okay,' he says.

With Amanda, his days off are spent travelling to B & S (Bachelor & Spinster) balls all over New South Wales and Victoria. He says Amanda recently hit two kangaroos in his ute going to the Swan Hill B & S. His work has always been outdoors, in shearing sheds, or doing general farmhand work such as fencing. **'A ute is essential. I couldn't do without it. A ute can take more of a pounding than a car. I love 'em.'**

He now wants to buy a LandCruiser — a bit flasher. Did he want to say anything about the ute? 'Yeah, it's FOR SALE.'

He plans to get his licence back in August.

A SWAGMAN & 3 UTES

I drove 815 kilometres to see Geoff Naylor, who lives on the Murray River banks in South Australia, only to find his camp deserted except for a big black dog, and a small bitch and her pup, just a few weeks old. So I jumped into the back of the ute, unrolled my swag, pulled my out computer and wrote up the previous day's interviews while I waited for Geoff to return.

He finally arrived home four hours later from a scrounging trip. Geoff is a retired 82-year-old cannery worker and strong union man. He is also a retired swagman. Walking the hundred-metre circle around the tin shed Geoff calls home, I shot three or four rolls of film. It is a treasure-trove of discarded bits 'n pieces.

Utes have played an important part in Geoff's life. He's had a 1955 Morris, three Holdens and three Valiants. Although second-hand, they were a luxury compared with the times he walked hundreds of miles while 'humping bluey' during the 1930s Depression.

Geoff's car 'graveyard' includes a number of cars and a van, as well as three utes, and

a collection of parts to keep any would-be enthusiast happy. But don't even think of trying to track Geoff down — he loves his privacy. He has lived without power, telephone, and mains water since the 1970s. Still, with his dogs and his big library, he's content to pass his days in primitive conditions but enjoying the peace and quiet.

Geoff remembers the little 1955 Morris: 'short on horsepower but strongly built — I often carried a ton of wood on it, but it managed to chug its way up the hills.

'I bought a Holden thinking I'd have more power, but it only carried half the load, and you'd have to hang onto the steering wheel all the way home — the stinking mongrel bastard of a thing.'

Of all his utes, he remembers the Morris as the best. He carried sawmill offcuts, and was always worried he'd blow the engine up. Geoff would rather a ute with heavy overload springs any day.

He likes people and he likes to help his mates. 'One for all and all for one,' he says. 'A ute is especially good for helping out mates.'

Once, a semitrailer rolled and spilled its load of wheat, so Geoff collected the wheat mixed with sand, loaded it up, took it home on his ute and sieved it all through an old wire bed-frame and mesh to get rid of the sand. One of Geoff's mates who owned fowls ended up with 21 bags of good wheat which lasted him a year.

Geoff's camp in the bush is often flooded by the waters of the Murray and his place becomes an island. He reckons the 4-wheel-drivers often churn up his track and he gets bogged. 'Once I was driving in the mud and slipping and sliding my way home. Next thing I know, the arse-end swerves and the ute ends up with me facing the direction of where I'd just come from.' He laughs heartily.

Geoff has memories of two T-model Fords. The first was during the Depression.

'In 1937 a family at Balcaldine, in Queensland, offered to sell me a T-model for £5. But of course as a swagman on the track I couldn't come up with the five quid. I wasn't afraid to work hard but I could never make enough. The ute would have been a help.

'Near Rockhampton I got a bag of cabbages for sixpence each and sold them for a shilling each. The local greengrocer was charging about 1s 9d, I think. He would have loved to kill me. If I'd got the ute, perhaps I could have worked and built myself up a bit. But it was not to be.'

Many years later and with another T-model Geoff recalls: 'I was driving a T-model as I had work near Adelaide. I had an unexpected tyre blow-out. I pulled the tyre off and rode on the rim for about eight miles before I could get it repaired. Now that was real steel. Imagine doing that with some of these lightweight wheels now.'

One of Geoff's last Valiant utes brings other memories.

'I'd only had it a month when a local copper chased me and pulled me up. He reckoned there was too much blue smoke coming from the vehicle. It cost me $400 to fix it up.

'In the end it just had too much rust in the chassis. It had gone that far, it couldn't be welded. So I had to dump it.'

Geoff's just turned 82 and is convinced he'll be around to see a hundred. He had his driving licence renewed recently after a medical examination: eyesight good, heart good, blood pressure 130 over 80.

Geoff says with a cheeky grin: 'Not bad for a youngster, eh?'

'Isn't it beautiful?' said Mrs Francis, describing her son's pale blue 1956 Austin A40 HiLite coupé utility (so named because of the unusual wrap-around rear window).

Dad has an Austin sedan, big brother, Adam, has one too, but I was here to see the ute. It is in pretty good condition for its age, and although it's been repainted, you can tell by looking it over that much of it is original. Its body was made by PMC in Sydney. It has a 1200 cc 'B' series motor. Martin Francis bought the ute in 1997 and is the third owner: the first was an old bloke in the bush, then a man who used it for advertising his tyre business.

'I think he just got it out of the factory first thing in the morning, parked it out the front, and then brought it back in at night,' says Martin. 'I knew about it even when I was still in high school, but he wouldn't sell it. Then I heard it was for sale, so I went and made an offer which he accepted. In any case, Martin doesn't seem to be in too much of a hurry to sell it. All the family are members of the Austin A40 Car Club of Australia. His brother, Adam, is President.

Dad owns a 1951 Austin Devon sedan, and big brother a 1953 Austin Somerset. Brother Adam 28, bought his when he was 16 and did it up so he could have a vehicle to put on the road with his 'L' plates when he turned 17. Dad got his about ten years ago. 'He grew up with cars of that era,' says Martin.

I had a good look at the ute while Martin went and got his dog Spod, the 11-year-old beagle–labrador cross. He was pleased to say 'G'day' to me, but was more interested in sniffing my camera case than having his photo taken. He loved it when he got to sit in the front seat.

Martin Francis 25, is a qualified gardener. He did his four-year apprenticeship with a city council and now works for a company that does all sorts of garden work in the private gardens of large homes in exclusive suburbs. 'We do maintenance work, turf laying, landscaping, planting ... we'd even clean the barbecue if we were asked!'

At the moment he drives an 1985 Corona to work and it's a bit of job getting his ute off his brother. Adam borrows it because his own Austin is playing up. 'But I try and get it back at weekends, though,' laughs Martin. The problem the three Austin owners have is that most parts aren't interchangable between the vehicles.

Martin's ute has about 64,000 miles on the clock; it is in good condition, and is 'a good practical vehicle and it has carried bulky tables, stuff for a garage sale, gear for club meetings — all chucked in the back. I normally try to drive it at least a couple of times a week.'

Probably more, once Adam gets his A70 working properly again.

The 'AUSSIE EARL'

The Rt Hon. Sir Keith Rous, 6th Earl of Stradbroke

In the world of big business I'd say he'd probably be seen as very astute, tough — perhaps even eccentric. Keith Rous likes to put his own individual touch on things in his life. They say you should take a person as you find them. I found him to be casual, straightforward and friendly. And nothing like you'd imagine of British aristocracy. He says that he and Countess Rosie are simply 'Mr and Mrs Rous'.

He describes himself as a property developer and investor in both commercial and residential markets. I visited him at one of his properties — Mt Fyans Black Angus in Western Victoria. Over an early morning breakfast with Keith, Rosie and about six of the fifteen children, we shared tea, toast and Vegemite as the kids prepared for school.

He may come from an illustrious English family background, educated at Harrow and all that, but he is not your usual 'silver spooner'. In fact, I found him probably more down-to-earth, more genuinely 'Australian' than many so-called 'true blue' Aussies.

He is very much a Republican. One of his

children designed their own Aussie flag which now proudly flies on some of the vehicles.

'Australia's been good to me. Very good. It was a breath of fresh air in my life. I was expelled from school. England just wasn't for me. I emigrated to Australia in 1957 when I was 19. I never returned to England for 20 years and then it was only to fight a court case to kick a cousin squatter off our property. It was finally resolved in 1983.

'The last time I was there was in 1992 although we travel a lot in Italy and France, where we own a place. As for England, I am absolutely not interested in returning.

'I am a Republican, a swinging voter, but usually Labor. And the titles don't mean that much to me, in fact I believe eventually they will be illegal or won't exist.'

The 'Aussie Earl' as he is known says hard work in Australia in the 1960s helped him. He found it easy to accumulate capital here in the 50s and 60s. 'I sold TVs and encyclopaedias and I had two part-time jobs and one fulltime job at one stage. I've always been a salesman. I tell my kids to "buy in doom, sell in boom."'

Real estate and farms have always been his main interest — and cars.

'In the 70s I always had very good cars, mainly American Lincoln Continentals and GTO Pontiacs. I got my first Rolls Royce in 1978 and have had about eight since. I'm always buying and selling great vehicles with great histories.'

And what about utes?

'I couldn't run the farm without utes. We have a Toyota twin-cab and a Toyota LandCruiser ute — and a Diahatsu on the

12,000 acres here, which runs 3000 Angus cattle and 4000 merino sheep.'

He bought 1050 hectares (2600 acres) at Mt Fyans nine years ago; it is now 4900 hectares (12,000 acres) comprising what was 11 farms. The homestead was built in 1882 and Keith and Rosie have completely refurbished and restored it to its original condition. The property is well known for its natural springs with one producing a million gallons of spring water per day.

Five of Keith's vehicles are in the National Motor Museum in Adelaide and I visited them a couple of days before our interview. Two are utes. The first is a 1967 Austin 1800 Mark II ute which came with a property Keith bought 'walk in–walk out'. BMC introduced the 1800 ute to compete with 'the big three'. It has a 1.8-litre 4-cylinder transverse front wheel drive engine, with hydrolastic suspension. In 1969, despite having the largest carrying capacity only 1739 were sold, compared to 15,061 Holdens, 7486 Fords and 3108 Valiants.

The other ute is very special — a 1956 S1 Bentley 'woodie' ute (see page 17). It was converted into a utility from the original sedan to carry the yachting gear of Charles Lloyd Jones (a member of the David Jones department store family). Only 3500 sedans were ever built. Keith's Bentley has

MT FYANS
BLACK ANGUS

a 4.9-litre straight 6 engine and 4-speed auto transmission.

'I bought it on a whim, just for fun. It wasn't working and it took John Vawser in Sydney, who I bought it from, about nine months to restore it for me. He did a great job. It's heavy in steering, but a beautiful ride. It was a daily work ute here on the farm, carrying everything from hay to 44-gallon drums of diesel fuel; it was always fully utilised.'

With a touch of Keith's personal stamp, all vehicles in the livery of Mt Fyans are bright yellow with dark green signage. 'We've never had a vehicle stolen since we adopted the colours. I can park in the centre of Sydney with keys in, no one steals them. It was different in the past.' His Mercedes sports is another colourful moving advertisement for his business. A bright yellow Rolls Royce Corniche convertible is also in the Museum.

Utes play an important part in life on the farm. And so do other vehicles. Yet another unique touch in the unusual life of the 'Aussie Earl' is his bus.

'The 28-seater bus I want to alter to a 7-seater for workers and convert the rest of it into a ute, then we can carry 44-gallon drums of fuel and other materials'.

Sounds crazy — but what a ute!

BIG BLACK MOTHER

'**I**t really is a nice ute to drive, it's just that it looks radical.'

There's a bit of consternation in the Jeffery household. With two growing daughters Ashlee 4, and Shannon 1, and a German short-haired pointer called Monika, there isn't enough room for them all in Carl and Marie's 'special built' ute.

Carl has reluctantly made the decision to sell the vehicle he and a mate, Mark Wheeler, built. He says that logically it is time to get a twin-cab ute, but knows he will never build another like the 'one-off' HK Holden body sitting on a LandCruiser chassis which is engineer approved. This a real mean-machine. It looks mean and sounds mean.

'A twin-cab,' laughs Marie, 'is a real family car. He is a considerate husband isn't he? We all love it, and I can understand his point of view though I know he really doesn't want to sell it. He shouldn't, we are proud of his achievement, it's a real head-turner, a great ute.'

Carl Jeffery is a qualified mechanic who has lived in Victoria's northern Wimmera district all his life. He lives for fishing and shooting. He built the 4-wheel-drive Holden in Mark's garage over a nine-month period, mainly at weekends or after work

when it didn't interfere with a shooting trip. He hunts pigs in New South Wales and decided he would have to have a special ute.

'I started looking through magazines, saw something I liked, sat down and did some measurements to see how things would fit. I bought a LandCruiser chassis and bits for $3500, then the Holden ute as a wreck that had only done 36,000 miles. It had been totally written off at the front. I bought it for $100. It was stored on top of a pile of old wrecks at a wrecking yard. I discussed it with an engineering company and kept them informed about what I wanted to do. It's cost nearly $10,000 to construct.'

The ute has a Holden 253 V8, dual fuel, power steering and air conditioning. Carl adds, '**It took a lot of work and a lot of beer.** We've had people chase us through the streets just to have a look at it. When it's sold I hope it leaves the district. I wouldn't like to see anyone else own it and drive around in it.'

BLACKSMITH

R obin Hatcher grew up as the son of a railway ganger and spent his childhood moving from town to town. He says his education stopped at a very young age. One of those days when he was at school happened to be a career day and he filled in a form for a blacksmith or fitter-and-turner apprenticeship. The next thing he knew he had passed an exam and was accepted as an apprentice blacksmith on Victorian Railways where he stayed for seven years.

That trade has become his life and he has worked for about 30 years as a blacksmith. He now lives off the beaten track in the hills on 5 hectares (12 acres) and enjoys the peace and quiet with the two ladies in his life, partner Margaret, and Lady the dog. There are other dogs also, Fleur, and the big old Bull Arab called Saab.

Lady won me over very quickly. She was a bit agitated when I

approached her and Robin said she didn't like people all that much. Once I got down to her level and let her approach me, sniff and check me out she relaxed and we became great mates, and I earned a big smooch. She loved having her photo taken.

Robin has a huge new shed in which he has a lifetime's collection of blacksmith's tools. Old steel-bending equipment, presses, drills, forges, anvils, tongs — you name it. His shed stores thousands of dollars worth of amazing equipment.

He drives a $450 1970 Toyota HiLux ute. A motor mechanic friend of the family was owed money from a customer. The person couldn't pay so gave the ute instead. Robin then bought the ute and it has proved to be a trusty workhorse.

'I carry heavy steel nearly 20 feet long on it. I built a special easy-to-install rack for it and I've just built a toolbox underneath as well. I'm in the process of building a special crane to lift steel on and off it. I saw one in a catalogue, so drew up my own.

'The ute is essential. Especially for my work. It's good for taking rubbish to the tip, carting bales of hay and more. I took the dog to the vet on the back of the ute the other day. Margaret has a little Suzuki;

have you ever tried to fit a Bull Arab dog into a Suzuki? I've had Holden tray-top ute before. It was made up of many Holdens, but mainly from an HQ. Nothing worked on it; it was terrible. I decided to sell it and the same day I put a FOR SALE sign on it, I had to get a ride home.'

Robin has that typical Aussie bush attitude when it comes to utes. The front of the crown of his hat wore out where it constantly rubbed on the roof of the Holden, so when he bought the HiLux he decided to adapt the ute so he could drive with his hat on. The solution?

'I ripped the bench seat out, lowered it by putting a sheet of chipboard to sit on and another for the back. I put some foam rubber on it and covered the lot with a seat cover. Now I can drive without wearing my hat out on the roof.'

Robin has a trade certificate allowing him to teach his trade at a TAFE College, which he finds amusing seeing he left school with little education. He showed me what he was building, and there is no doubt that Robin is a talented man when it comes to using steel. He is very inventive, and working with iron and steel comes naturally to him. He says he is not the best tradesman, but that was understatement. He says casually, 'I am proud of my craft and what I do.' He makes all sorts of items: garden arches, banister rails, hinges, pad bolts, gates. 'You name it, I'll make it, if it is made of metal.'

Before Robin bought the Toyota ute it had been lying in a shed unused for six years, and before that was in a caravan park.

'It's the best ute I've owned, but the arse is too light. I went looking for the dogs one day in it, but it couldn't get up the hills. **It won't be sold until I either run it into the ground or I win Tattslotto. Then I'll by a King Cab.'**

'I've got two ex-wives, ten kids, 37 grand-kids, and four great-grand-kids that I know of. I'm an arrogant, self-reliant, down-to-earth old bastard that believes laws are for me as much as any-one else, and I should not be singled out. Everyone should be free to speak and live free. I've lived a bloody hard life but it's been educational and a rich, rewarding one.

'I love my kids, never took shit from any-one, never lay down. I'll help anybody, but don't like to be shit on.'

'Bluey' is an Australian bush character, never short of words when it comes to telling one of his many yarns. He says he was discharged medically unfit from the RAAF in 1944. He has worked around cars ever since as a mechanic and been involved in garages, caryards, and wrecking yards ever since. He was doing up three cars when I visited him on his 15 hectares (38 acres). His family is now two little dogs Boomer and Patches, 10 horses, 17 goats and 5 cows.

Bluey says his 40-year collection of 322 cars was taken by the local council, over a three-day period, to Sims scrapmetal yard. He is now suing the council. He says he's been in and out of courts suing the council, electricity company, banks and more over the years.

'I refuse to lie down to people who should have known better.'

Bluey has been in trouble more than once; 'got put in a mental home in Goulburn about 1973, but the RSL came and got me out. I also did two years in Victoria's Morwell River Prison Farm years ago.

'During the Depression as a kid I carried a heavy suitcase full of clothes for my father. We hawked them around to anyone who'd buy them. My mother made the clothes on a sewing machine that I still have in the house.

'Things aren't always easy. I was one of nine. My father was machine-gunned during World War II. I joined up to avenge his death — was going to kill them all, but of course I didn't.

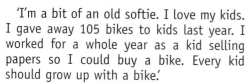

'I'm a bit of an old softie. I love my kids. I gave away 105 bikes to kids last year. I worked for a whole year as a kid selling papers so I could buy a bike. Every kid should grow up with a bike.'

There are about 30 sorts of bikes awaiting fixing in his yard. 'I'm good at scrounging.' Amongst Bluey's junk are about a dozen cars of various makes, including two utes: one a Mazda with the motor out, and the other his little 1982 blue Datsun 1200 with 469,589 kilometres on the speedo.

'It's been a mighty ute! I got it from a city council derelict yard for $150, seven years ago. It's on its third motor now. It's been everywhere — and I mean everywhere.

'I've carried up to a ton on it. Boy, it was nearly on the ground!

'It's definitely not up for sale, although I've had many people wanting to buy it.'

Bluey's ute will go on for a few more years if he has anything to do with it. Bluey is one of those independent Aussie characters that sticks to his guns no matter what. His yarns are also his beliefs. He says

his brother died of cancer, and he also was also diagnosed with cancer in 1981. He was given three months to live, has had major surgery three times and is still going.

As he lit another fag, he said, 'I just don't believe in lying down — I'll reach a hundred if I can manage it!'

BRONTE

When Brenton Rohrlach decided to get married he sold his ute to the youngest of four brothers, Bronte.

Bronte 24, is a fitter and turner with GMH in Adelaide, and I chased his ute down the road in the Adelaide hills where he lives with partner Anthea. He used to get pulled up by police in his mean ute, but now they pick on Commodore drivers instead.

His ute is a burgundy 1976 HX Holden front half, with a black 1954 Chevy rear. It has a WB Statesman grille, a HZ Statesman interior, sunroof, 9-inch diff, all discs and runs a HP308 HQ Monaro motor, with 15 x 10 inch tyres on the rear, and 15 x 7s on the front. It runs on straight gas, no petrol.

'It was originally a farm ute, then converted. My brother did the body work. I did up the motor. It's very comfortable. All it needs now is a new paint job.'

'BURKIE'–
BOGGED, BURNT & SHOT

'**T**hat's not rust down the side of the ute, it's just a nasty beer stain.'

Brian Burke is famous for his dry humour and lightning-fast wit. I'd describe him as a 'real' Australian, a hard yakka man, full of the larrikin spirit, a strong but lovable bush dag. He's a big man with a laugh to match. He is a second-generation farmer: his family have been on the same bit of dirt for the past 60 years, all 500 hectares (1200 acres) of it south of Shepparton in the Goulburn Valley, Victoria. Home is with wife Bernadette; they have four sons — Alistair 25, Chris 22, Damian 19 and young Shane 17.

Brian's fame had reached me before we finally met. When inviting me for a beer, he said there were some 'toy' beers (his way of referring to light ale); he doesn't mind a drop, old 'Burkie'. He said that '**drinking toy beers is like dancing with your sister. You know it isn't going to lead anywhere.**' A great loud laugh followed.

'Burkie' learnt to drive on the farm when he was six years old. 'All my kids learnt when they were six or seven. By the age of eight, I used to drive the old 1952 Bedford down to catch the school bus. I had a few old bags piled on the seat so I could see over the steering wheel, and I had to slide down off the seat to reach the pedals and shove it into whatever gear I could find. I

remember the doors used to open from the front out, opposite to the new ones. It'd do 60 mph flat out. It was a mighty old ute!

'When I was about 11 I tried to take the bridge near home flat out, but lost control, slid sideways and ended up across the bridge. It took me half an hour to get it going again. No damage done.' That laugh bellowed out once again. Burkie loves utes but they get a hard time. He's had a few

Holden utes: two FJs, an EH, an FC, an HQ. He says the two VC Valiant utes were the best.

'**The VCs were tough. I hit a roo and killed it and all it did was buff the paint up a bit!**

'We've always had at least two utes. Dad's 84 now and still drives a Subaru. He's also had an FJ, FC, EH, VC and XY. The Subaru's a good old-bloke's ute.'

Burkie has had two utes which are remembered by nearly everyone in the

district. He showed me one in a shed, an HQ Holden called 'Bill' (see page 28).

'Bill Anderson was an old cocky. I saw his old HQ in a caryard and thought it'd be a good buy; he'd always looked after it on the farm, so I bought it. Everyone started to call it "Bill", even old Bill Anderson himself used to say, "How's Bill? You looking after him?"'

'Bill' has seen better days. He has worked hard, still goes but you can't open the driver's door, the seat is stuffed, and rust *rigor mortis* is settling in.

I'd heard of another ute and Burkie was happy to recall it as he got stuck into a pizza and coffee — his lunch at 4.30 p.m.

'It was an old 72 Falcon ute; it had had a real hard life. Once I was driving in Shepparton when a nasty copper pulled me over for no seat-belt. He walked around the ute shaking his head, came back to the driver's door and said: "This thing is a heap of shit!" I looked at him and replied: "so would you be if you'd been bogged, burnt and shot!"

'He looked and said: "Get whatever you have to, then get this thing out of here and piss off home along the back roads." He let me off. I couldn't believe it, he was known as the hardest cop in town.'

Burkie roared laughing and then he ex-

plained: 'The old ute had been bogged a fair bit over the years and was always bloody dirty. The old man never worried about keeping utes clean but he always did with the motors. He was washing the motor down with about half a gallon of petrol. Anyway, my son turned the ignition on, Dad dropped a tool which arced the coil and a spark set off the petrol. The whole lot whooshed up and the bloody thing caught fire.

'I was inside the house and could hear the old man screaming but by the time I got out **the fire had burnt the whole front off the ute. Flames everywhere. All the plastic was one big lump on the ground.** Anyway, it was insured and it was rebuilt.

'We'd just got it back from the panel-beaters and Dad was off down the paddock shooting cockatoos in it. He crouched down in the seat to hide, sat up and snapped the double-barrel shotgun closed and the gun went off and blew a hole in the door. It was a small hole on the inside but on the out-side it blew the whole door apart, just

peeled it back. I said to the old man, "You stupid old bugger, you must have had your finger on the trigger." "No," he said, "it just went off."

'I sat in the ute and loaded the gun again, trying to re-enact the scene. I snapped the gun closed. Boom. It went off all right. I blew all the bloody rubber trim around the door to hell. We thought we'd better put that gun away after that!

'So it had a hard life, that ute. Bogged, burnt — and shot!

'Dogs? Aw yeah. They own the utes.' We went for a short drive in Burkie's working ute, an XD (see page 29). The seat is stuffed in it too. Dust 'n rust is the colour.

'Gees, I just cleaned it out the other day,' he said as I photographed the interior. On the oily rear tray the three dogs jumped in with glee, tails wagging, over the moon to be going for a drive. There was Tip ('got a small white tip on his tail, he's a fighter'), Tuckie ('one of my sons played for Hawthorn, we reckon Tuck's a cunning bas-tard') and Morgan ('he's a bloody mad young pup, so we named him after Mad Dan Morgan, the bushranger').

'The dogs reckon the ute is more important than a kennel. In fact Tip doesn't have a kennel, she sleeps in the ute!'

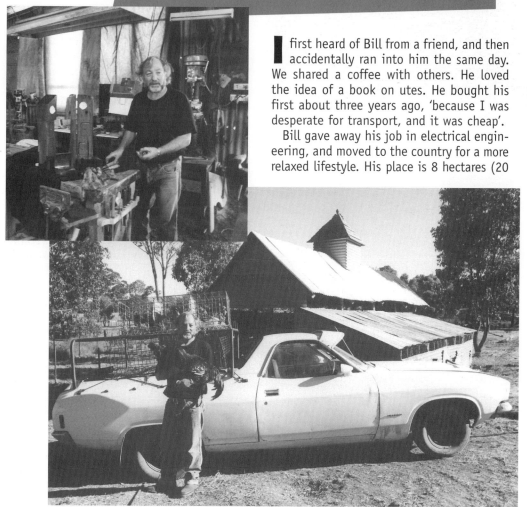

I first heard of Bill from a friend, and then accidentally ran into him the same day. We shared a coffee with others. He loved the idea of a book on utes. He bought his first about three years ago, 'because I was desperate for transport, and it was cheap'.

Bill gave away his job in electrical engineering, and moved to the country for a more relaxed lifestyle. His place is 8 hectares (20 acres) and off the beaten track. The ute is used to carry hay, chook food and loads of timber, mainly redgum for his woodwork business. He tries to keep six months reserve of raw material collected from all over the place which he turns into saleable items.

His XB Falcon ute may have been cheap, but it has turned out to be a reliable workhorse. For $600 and a new gearbox, it has motored on working hard every day.

The ute was bought 'out of the local rag'. It had been in the one family for about fifteen years, passed down from father to son to daughter. 'It's simple and solid, it's old and it's been good, except for the heavy steering and turning circle. It is now slowly dying of cancer (rust *rigor mortis*). It'll die. I'm currently looking for another vehicle, probably an XY Falcon ute.'

Bill has a hobby and a joy. He's a chookologist. And what is that? He breeds rare Speckled, Silver and Red Sussex, Aracauna, and Faverolle poultry. He is interested in genetics and breeding, and has been buying, selling and swapping for about seven years. The ute is in great demand when it comes to transporting cages of his fancy chooks. These free-range chooks were difficult to round up for a photo-shoot!

Something tells me there will be another ute very shortly.

CHRIS, RUSTY & THE RIG

Chris Pianto is a survivor. He was a victim of childhood sexual abuse by a paedophile school-teacher. Chris is one of those blokes that has had turmoil and tragedy shape his life. He has shown great courage and strength, which has helped him go forward. He is a quiet 39-year-old bloke who is as creative now as any artist. It wasn't always that way. In fact, self-destruction was more his way when he was younger.

After a ten-year ordeal Chris finally won his legal battle. He stuck to his beliefs and challenged the system and fought for justice. It was not without its price. He shot himself in the leg to expose the paedophile and as a protest against the justice system. At first denied a court hearing he gained national media attention and was at last able to get his day in court, with the support of other victims.

'Not only have I survived the dramas, I am normal. I have never given men or little boys a second look, but I certainly don't mind sharing a swag in the back of the ute with a nice woman! I'm a normal red-blooded Aussie male.'

Today life is much quieter and more stable for Chris. He is building his own home. He left school at 16 with little education, and three days later started as

a fencing contractor, which he continued doing for ten years.

At 27, Chris began labouring for a brick-layer. He liked being in the building industry, so spent two years in plumbing, and then in various trades such as concreting, carpentry and roofing. 'I'm no good with electrics though, I'd probably kill myself.' His enthusiasm for learning has stood him in good stead.

He is now a self-employed 'brickie'. He doesn't have an answering machine, he took his phone number off his ute, he doesn't advertise or display any signs, but he has more work than he can handle. He recently had to knock back four housing projects.

He is as particular with his ute and cement-mixer as he is with his work. One day a man stopped him in the street and asked, 'Are you a bricklayer?' When he replied he was, the man said, 'If your work

is as good as the way you look after your gear, you've got a job.

'He thought that the cement-mixer was too good to use! I must admit I hate to see cement on it, and I always make sure it's kept clean.' Chris gives a bit of a laugh. 'Actually, a lot of people ask if my work is as good as I keep my ute and mixer. It's a great advertisement.

'I'm probably a bit too fussy with the mixer, but it sits so well behind the ute. At 100 kilometres it never moves. I designed it myself, and built it to meet registration's requirements. I bought the mixer brand-new four years ago, ripped the pissy little axle and wheels off, built the trailer for it. A mate of mine had a two-pack of black spray-paint, and said he'd spray it to match the ute. Another mate liked it so much he gave me the two mag wheels to match.'

A real eyecatcher too is the black 1974 HJ Holden ute with a 308 V8 and lots of chrome. 'It's done about 300,000 kilometres, but the speedo hasn't worked since I got it about eight years ago. I just work off the tacho. I bought it for $4700 privately. It was pretty stuffed, with only a 253 motor, bits falling off it and a stuffed gearbox. I painted it from white to black, dropped in the 308, took the Aussie 4-speed box out, and put in a Ford toploader,

LSD, new mags, twin exhaust stacks, truck mirrors, a GTS dash and re-chromed it all. I made the grille myself. All up I've spent about $10,000.

'It's a work ute, so I don't worry about the interior much. In winter you have muddy boots and more, so it's not worth it. But I do like to keep the exterior looking good.

'I'll never sell it, I will just keep the maintenance up to it. Even if I did sell it, I'd only go and buy one the same, so why bother? I'll look after this one.'

Chris does a lot of shooting and fishing with his mates on the Murray River near Kyalite, New South Wales. 'I take the odd sheila out in it, do a bit of cruising, Sunday drives, whatever you want to call it.'

There is another permanent addition to Chris's ute. It is 18-month-old, pure-bred german shepherd Rusty. He is a seven-day-a-week dog for Chris and he goes everywhere with him. 'He won't let me get away. If he thinks I'm off, he's in the back before I can get in the front.'

This was demonstrated when Chris went to move the ute. Rusty was on board in a flash. One of most intelligent, gentle, good-natured and lovable shepherds I've ever

met, Chris has trained Rusty exceptionally well. We had a great time.

Chris's rig is well known around his sea-side township. Apart from its good looks, there is something else about the ute that attracts attention. A couple of his mates were always giving him that well-known sign 'the finger' (also known as 'the bird'). One day he backed down their driveway as they sat having a beer. They wondered what he was doing. Chris had attached a small rope inside the cabin, leading underneath the rear tray to a spring-loaded plastic 'arm'. When the rope is pulled, up the arm pops, finger extended. His mates admit they could never top that.

'Often the kids in town yell out, "show ya' finger!" so up it goes. I've only used it three times in anger at other drivers, but it's great to watch the looks on the faces of people in the rearview mirror. I've done it to cops and none have ever worried, usually they just laugh. I don't think they could do anything about it anyway. It's a bit of a laugh, and you gotta laugh, don't you?'

The paedophile school-teacher who caused much pain to many lives is now serving a minimum five-year gaol sentence. In an ironic twist, one of the man's childhood victims is now a gaol warder.

Chris has not only survived. He is able to help others now. He believes in educating children of the dangers, and hopes that his story may help others and show that it is nothing to be ashamed of.

'I am no longer a victim. I am free to enjoy life.'

'I've always wanted to own a ute, they're so practical.'

Twenty-four-year-old Claire Edwards has always loved, driven and owned all sorts of cars. She grew up surrounded by vehicles. Both her father and brother have been truck drivers, and racing waterskiers, so Claire heard all about engines from a young age.

Her father now drives a dual wheel Chev ute imported from the USA. After about his 35th car, Claire's Mum said the next one had better make money, so he now operates Hertz Rent-A-Car in Mildura, Victoria.

Although Claire works as a sales clerk for another company, she has ferried many cars from city locations for her father's business.

Her trips have taken her to Adelaide, Melbourne and Sydney.

Claire is a single mum, with young Jay just started school this year. She says she has great family support from both her families.

We first met when I saw a pale yellow ute driving across a wind- and dust-swept plain on my way back from the South Australia–Victoria border. Claire was headed in the opposite direction.

I did a U-turn, chased the vehicle and pulled her over. I explained about my book on utes, and her enthusiasm was instant. She was travelling to South Australia with a friend Ian Bish, a long-haul truck driver. Claire and Ian have been mates since primary school and are now flatmates.

They were more than happy to show off the yellow 1987 XF Falcon. We took a few photos before rain brought it to an end, and she drove off down the track again and me in the opposite direction. We talked again a few days later.

Claire has been waterskiing since she was four years old, but competition waterskiing 'takes the fun out of it', and she is content to plump for just the fun. She loves to go camping on the Murray River, and she also loves horses.

Her first car was a 'brilliant' 1969 fastback Volkswagen, then a 1985 Camira stationwagon ('a dud, no one should buy one'), a Cortina, a VL Valiant sedan, and a Suzuki Swift. The yellow XF is a good example of turning a standard ute into something eye-

catching without spending vast amounts of money. It is a stunning vehicle with nothing more than 14-inch dragway wheels, tinted glass, and it has been lowered — all for less than $1500.

Claire sees a ute as being practical. It has

'the look' and, from her point of view, a ute certainly is 'a bit different for a female driver'.

This woman of the 1990s in her Doc Marten boots and swags in the back of the ute seems to know where she's heading.

Col Minogue comes from Goulburn, New South Wales. He started work on the railways as a 15-year-old, riding 38 kilometres a night on his bike as a call boy (waking train drivers up to get them to work). He later worked as a fettler and in various railway sections — signal points maintenance, platforms, locos, sectionman, ganger on bridges, and buildings. You name it, he did it. Col retired from the railways after 43 years. When we met he was on tour chasing family history in cemeteries across New South Wales.

His 1996 Toyota twin-cab HiLux is a great set up. It has a two-way hydraulic tip-tray, a removable canvas canopy, and tows a 'special built' trailer which was the original tray rear of the ute. The conversion was begun when someone smashed into the ute and did $1200 worth of damage. He converted the back into an aluminum dropside tip-tray, and rebuilt the rear as a separate trailer with the help of a mate that he works for as 'gopher'. His mate owns an engineering workshop, so Col got all the work done for nothing in lieu of pay.

He is brilliantly set for travel with his little friendly ten-month-old Jack Russell terrier 'Mac'. The ute has a bed, a three-way fridge, stove, mobile phone, TV, and all the camping gear you'd ever need. The trailer carries 85 litres of rainwater, tools, tent and anything else Col needs to be mobile, free and self-sufficient. He maintains it like new.

We parted company after a great yarn; he was off to another cemetery.

'I'll be home in a week or six months', he laughed as we bid our goodbyes, he in one direction, myself in the other. Mac was on Col's knees looking out the window, wagging his farewell.

'CONCRETE CHARLIE'

'**C**harlie went all his life and never owned anything except a ute.'

Charlie Gill was a concreter in Casterton, Victoria. He did all sorts of work ranging from building sheep dips, chimneys, guttering, setting in stoves, fencing. He even built the Sacred Heart Primary School. His son Tony was a concreter, as is grandson 'young Tony'. The three generations built the local town swimming pool in the 1960s.

Charlie was a real character. He was known locally as 'Concrete Charlie' or 'Monday Charlie'. Asked when he was going to start a new job, it was always 'Be there Monday'. His son was 'Tuesday Tony'.

Concrete Charlie never owned anything but a ute. He owned a number over the years — a Rugby, a T-model, two Chevs (one he bought new in 1936) and his last an FX Holden, which never went out of second gear.

'**He was never really taught how to drive, so second gear was good enough, and besides his ute was always loaded down with gear, such as the hand-mixer for cement work.**'

The utes were also loaded down with kids. He and his wife Annie (Hannah) had six kids, who rode in the back whenever they went anywhere, including 60-odd miles down to Portland. You don't see people travelling in the back of utes like that any more.

I found out about Charlie through his granddaughter Bernice. She spoke with many members of her scattered family and a picture of the man slowly unfolded. Bernie's own memories of her grandfather show a natural warmth and love for the old fella.

'I remember as kids, we'd be in the lounge in front of the fire playing, building houses out of playing-cards. Nana was always in the kitchen waiting for him to come home from the pub; dinner would be ready.

'**He would never eat anything out of a can, and never anything cooked in an electric frypan.**

'We would hear the flywire screen door open and we'd yell "Pop's home" as he came in down the hallway whistling. He'd plunge his hands into his overall pockets and pull out fists full of change, pennies, three-pennies, sixpennies, whatever he had. He'd throw it down the hall with all us kids chasing after the scattering money. "There you are darlings." He always called us darling.

'He was never seen without a smile. He was happy with his lot in life.'

Bernie remembers her grandfather's FX

Holden ute. 'I'd be squashed in between Pop who wasn't very big, and Nana who was a "generously proportioned woman".'

Concrete Charlie was a generous man. He took people at their word. Although he ran his own business he never kept account books. His quote for a job was always worked out in his head on the spot. He'd squat down, work it out and it would be written on the back of his hand, on a scrap of paper or the back of a matchbox. He never sent out a bill.

He'd get paid in the pub or down the street. He trusted people. Some may have taken advantage of him, but he survived. During the Depression when people couldn't pay, he'd tell them to pay him when they could. Often he would be paid not in cash, but with a side of lamb or other foodstuffs. He never asked for money.

Bernie tells a family joke: 'Two old ladies asked Charlie to do a footpath for their home. He said he'd be back "next Monday" and in the meantime they should move the bags of cement each day to stop them from going hard. Six months later they bailed him up in the street; they'd been moving the bags all that time.'

Charlie's son Tony courted his wife Ruby in the old Chev ute. He also worked as a concreter all his life. Grandson 'young Tony'

went on to buy a hardware store. Tony also owned a Zephyr ute at one stage, and certainly knew how to move it through the gears!

Charlie never had a sick day in his entire working life. When he finally died in 1966, aged 72, the doctor said Charlie 'just wore out'. He is buried in the Casterton Cemetery.

Kevin Dalton is a busy man. He is based in Darwin, but is a hard man to catch as he works right across the Top End, from Cape York in Queensland to Broome in the north of Western Australia.

I tracked him down by swapping mobile messages and we finally ended up in the same place, at the same time, for a photoshoot, a quick look over his rig and a bit of a yarn.

We shared a cold drink from the little fridge in the rig. We decided we'd leave the full interview until we both had more time. It took us many more mobile phone calls, answering-machine messages, faxes back and forth before we finally caught up with each other again.

Kevin works in the building and construction business. He builds houses, sheds, and other buildings in Aboriginal communities and other remote outback areas.

'Ninety per cent of the time, I am in the remote parts so you need a vehicle that is home, workshop and warehouse, as well as being a "go-anywhere" tow vehicle.'

The Toyota LandCruiser meets his needs really well and he has done all the work on it himself. It is a 'special built' vehicle for a special job. The rear tray is loaded with power tools, saws, ladders and every other material a builder needs.

'It was a 1991 ute when I purchased it in 1992. First I decided to rebuild it to suit my needs. Having an engineering background, I set about lengthening the chassis so it could carry more load. I fitted a lazy axle and another set of wheels using a Hayman Reece stretch pack conversion. This gives it 45 per cent more carrying capacity.

'It also increases braking efficiency because you have six wheel-brakes and handling is improved

with the added rubber on the road. In some cases it is better in the rough, for example, in a bog the rear wheels seem to hold it up long enough for the front wheels to pull it through.

'Having lengthened the chassis, I enclosed the tray with roller shutter doors on each side, then made and fitted an outback dust-proof 'dog box' between the cabin and the tray. This is most useful to keep food, hang up clothes, and lock valuables in. I included a fridge in it to keep beer and food cold.'

Kevin's rig turns heads wherever he goes; the quality of his workmanship is outstanding. Kevin has made his vehicle self-reliant out bush. It's a great rig for the last frontier of our vast continent.

'I have fitted a second fuel tank under the dog box, which allows me to travel 1200 kilometres between fuel stops. The 4-litre engine with ARB turbocharger gives ample power for towing caravans and other heavy loads. Driving comfort comes from Recaro seats which keep my back in trim in the rough country. The ute also has a dual electrical system and a heavy duty radiator.

'The 8000-lb winch on the front has never had to pull her out of trouble, but has been used extensively to pull and tow materials and gear in very remote areas. The ute cost me $22,000, and I've spent another $8,500 so far, excluding my labour.'

If you see Kevin outback, pull over and stop for a cuppa.

EMU EATERS B&S
Beer, Bundy & A Mug on a Rope

The atmosphere is amazing. I was welcomed with warmth and generosity. Most were nearly half my age, but it didn't seem to matter. Choking dust can't tame the free spirits of country folk when they come to a B & S (Bachelor & Spinster) ball. Jackaroos, teachers, farmers, uni students, office workers — Aussies having fun.

There's more cowdie hats than in a John Wayne movie, more dust than when the Cavalry chases the Apaches — but there's no shooting in the desolate paddock, just the roaring bark of V8 utes doing circle work in the dust.

'Better cover ya' beer up mate,' says a jackaroo as he throws a huge mitt over the neck of his stubby. Bugger the beer, what about my $2000 worth of camera gear? No one gives a shit about the dust, you just sit in it and what doesn't fill every nook and cranny will blow on through. Your only worry is if the dust makes the beer or bundy rum gritty.

One jillaroo summed it up, 'Sometimes I might be out bush working for three weeks and hardly see anyone until I catch up with all my mates at a B & S. This is the only place we get to come together and let our hair down.'

It's party time in the bush, and country blokes and sheilas know how to have a good time. The only real idiots I meet are from Melbourne. There are plenty of drunks, but no sign of drugs. People just want to have fun. All are sick to death of being harassed by cops. They call 'em the Funbusters. Sometimes the language would curl your grandmother's hair.

'Isn't it better that we are out here on private property off the roads letting our hair down. There's no drugs. The idiots that get drunk and hit the roads, end up in fights, doing damage or whatever, they don't belong here. There's always a small group that stuff it up for others. We all look after each other.'

The thing I noticed was the respect everyone had for each other. Not one fight broke out. By the end of the night I'm sure there were a whole heap of drunks, swags being shared and condoms all the go (I left early to protect my virginity and my 21-year marriage!). In New South Wales, the authorities conduct breath-testing at B & S balls; if you try to leave you are tested and sent back until you are sober. It's a good idea and some Victorian B & S balls are now doing the same.

'You stay here until you sober up, only the idiots drive out of here. Besides if you stay you might get a root.'

For an old bloke of 47, I had a ball. I left with a positive feeling that this was the place for them. Sure, there've been some problems at B & S balls over the years, but how many young blokes and sheilas have we lost on our roads over the years?

Instead of the wowsers trying to close B & S balls down, they should be encouraging them, helping keep their kids alive and enjoying life. B & S balls should be fairly policed, and well organised. Idiots should not be welcome — and if you drink and drive you are a bloody idiot.

The B & S also brings in big money for small rural places. All sorts of people benefit. Security staff, caterers, butchers, bands, milk bars, publicans, and many other suppliers. Donations from B & S balls go to local sporting groups, fire brigades, SES (State Emergency Service) and bush nursing centres. One town estimated a B & S ball weekend was worth $550,000 to its community.

Many of the kids were well educated (and yes, some were bloody morons), some were studying for degrees, and, surprisingly for the country, every single person I met, male or female, had a job.

The mug. It was tied to the wrist, or draped around the neck and some tied to the belt of their jeans. 'What's the mug on the end of the rope for?' I stupidly asked a bloke. He looked at me as if I must be from another planet. 'Ya' get pissed and fall down and a glass would get broken, and so you don't lose ya' drinking mug, mate, when you come out of your drunken stupor, mate.' Yeah, right. Pretty well a non-drinker now, if I'd drunk every beer offered, I wouldn't have woken up for a week.

Dust. Blimey — I've only got one and a half lungs — it took me a week to breathe again. My new ute will never be the same. But I got to the Emu Eaters B & S Ball.

I wouldn't have missed it for quids!

Proudly Supported
By
"EMU EATERS
B & S. BALL"

HOT DOGS $2.50
PLAIN - $3.50
THE LOT
BAKED SPUDS - $4.50
DIM SIMS - $.50

HOT CHIPS
ICY POLES
CONDOMS
BEROCCA
PANADOL
SEX

Bundaberg
RUM

15 dead Emus, 90 howling Dingoes, 3 barking Dogs ... & a Holden Rodeo

I rang Bruce Jacobs at 'The Dingo Farm' near Castlemaine, Victoria, to see if he currently had a ute. As a result of the call, I dropped everything and a few hours later we were seated on a large log talking politics, dingoes and people we knew — surrounded by 90 dingoes led by Wingnut, three dogs Ralph, Growzer and Skiddy, and a few bush blowflies.

Bruce is probably the most famous breeder of Australia's native dog. You may remember him making front-page headlines when six of his dingoes were shot in a raid by the RSPCA and the Department of Conservation some years back.

Bruce took on the challenge, the ombudsman investigated, Bruce won and helped to change the laws in Victoria about dingoes. You can legally keep them now. Today his farm is internationally recognised.

Dingoes came and sniffed and stopped for a pat as I interviewed Bruce. He has an amazing eye for detail and can name any dog that wanders by, its parents and other details. He casually made a 'roll-yer-own', as he yarned about his beloved dingoes.

He gave me some family background — divorced, one son, but when I asked what year he was born (which is my polite way of asking 'how old are you anyway?'), he gave a bit of a laugh and a 'get stuffed'.

His father once owned the Gaffney's Creek pub in the mountains of eastern Victoria. It was there at the age of four that Bruce had his first dingo pup, straight from the mountains. He's had dingoes ever since. The pub was burnt down a few years ago when two prison escapees and a female warder took refuge there before being caught. But that's another story.

Bruce interrupted his yarn, 'that sounds like the ute coming now.' Before I knew it he was up and off to open the gate with

most of the dingoes headed in the same direction. One howling dingo set off 89 others. An amazing sight unfolded. A white Holden Rodeo ute made its way down the hill, dingoes following and howling. Those in pens were as excited as the ones on the loose. They could smell blood and crowded around the ute.

Removing the tarp revealed the bloody carcases of 15 dead emus. What better way to preserve the natural instinct of the native dog than to allow it to feed on the huge bird from the Australian bush? The emu meat was destined to be burnt otherwise. The birds had been stripped of their feathers and were ready to butcher. Blood ran from the aluminium tray of the ute into a large plastic tub underneath. Excited dingoes sensed the kill and were trying to get to the emus.

A large, razor-sharp butcher's knife soon made short work of the emus. With a wheelbarrow full, Bruce headed off to the pens to give the dogs and bitches a leg or two to rip apart, leaving offsider, Miles, in leather apron, to carry on butchering.

Dingoes were everywhere; large packs moved in, fights broke out, dominant dogs overpowering the submissive ones. There's a strong pecking order in a dingo pack and they all know who is boss. Today Wingnut is boss dog, and Mabo and his pure white mob are penned up. Their turn to roam the 5½ hectares (14 acres) will come tomorrow.

I couldn't take photos fast enough, I didn't know which way to turn. The frenzy was on in full earnest.

I interviewed Miles as he continued to butcher.

MILES
Vegetarian Gardener turned Emu Butcher & Dingo Minder

'**I** bought the ute off a couple of Mormons, I knew they wouldn't lie or they'd go straight to hell, so I knew I had them and got it at the right price.

'I was a bit pissed off that they'd taken the radio out. They weren't allowed to listen to those heathen sort of things.'

Like his boss, Miles is a bit of a bush larrikin. Down-to-earth. He was covered in blood and hosing down the back of his ute as we talked. He was born at Warrandyte in 1970. I'd already met his mate Skiddy, a blue heeler with a black patch over one eye.

Miles commented on how he had been a vegetarian, motorbike riding, permaculture glasshouse gardener before he started at the Dingo Farm less than 12 months ago. He never thought he'd end up doing what he's doing.

He finds it interesting. With TV crews from all over the world coming to film, tourists by the busload and wandering authors, life on the Dingo Farm is never too dull.

The Holden Rodeo 2600 ute was bought secondhand locally in 1991. He's still single and the ute and his dog are constant companions. He does a lot of camping, but is quick to add he never sleeps in the back anymore.

He has used the ute for all sorts of things, first in his gardening work and now at the Dingo Farm. It has carried unusual cargo including 850 kg solar-panel inverters to a lighthouse in far east Gippsland.

He gives a laugh and relates how it also carried a big papier-mâché giraffe to advertise a community spring dance with an African theme. On another occasion it was dressed up for a Fringe Festival Parade with people on the back throwing lollies to the crowd.

Soon to be installed on the front bar are landing lights from a Fokker Friendship aeroplane which a mate is tracking down for him. The ute runs on a 100-litre gas tank and Miles says he's been 4-wheel driving in it, but 'doesn't recommend it' as you should have a 4-wheel drive!

With the clean-up of the ute and Miles finished, it is the end of the day for him. He's off to volleyball practice. We shake hands and he's gone, with Skiddy well fed and seated beside him.

Later, Bruce and I sit on the log again quietly yarning, the dingoes have settled except for the occasional 'domestic' between two of them. The sun is low in the sky, throwing long shadows through the ironbarks. Bruce is dragging on another

roll-yer-own ciggy. He talks about his work with scientists from various parts of the world and the dingo research that is currently being carried out. He gives an unusual summary account of breeding.

'They mate in the autumn and pup in the winter. The hormones really get switched on and by Anzac Day it's like a 'canine brothel' around here, but by 22 June the hormones are off and everyone's happy and settled again.'

Finally he shook hands strongly, reminding me to check out his Dingo Farm website, as yet another phone call rang out in the background.

I gave Growzer and Ralph, the 12-week old pup, a goodbye pat and said so long to Wingnut, Mabo, Mungo, Tiger, Mutton, Zac, Chook, Hazel, Smiley, Bungy, Livingstone, Laura, Kelly, Koori, Ned, Snowy, Buffy, Happy, Sweetie, Boof, Cloud, Crazy, and the rest I forget. As I headed up the hill, the dingoes gave me one last look. I pushed the padlock shut on the huge gates and turned to admire the view down the small valley to the dam where some dingoes were swimming. The iron-bark forest was quiet. A horse and donkey stood at the electric fence and silently watched me.

I drove along the dirt track back to the bitumen and headed for home. I had learned a lot this day — of dingoes, DNA, breeding and politics. An interesting man, the Dingo Man, and his workmate Miles, and their wonderful dingoes and dogs. Call in on the Dingo Farm one day and say 'G'day'. Tell 'em the Uteman sent you. I'm sure you'll be made welcome.

Great name, eh? Rolls off the tongue like a moviestar's name — Errol Flynn, Clark Gable ... I saw Mark Moretta's dark blue XF Falcon ute parked outside a builder's factory. It looked great, so I went in and made myself known. Mark Moretta describes people who drive utes as 'real men or women, not wannabes, just down-to-earth and genuine, especially in the country'. This is how Mark struck me.

He is passionate about utes, and owns two. He wouldn't be comfortable in anything else. His old red 1979 XD Falcon has now done 300,000 kilometres of bloody hard yakka. Mark is a carpenter by trade and has spent some time in country New South Wales and north-east Victoria building homes.

For these trips the XD was heavily loaded down with ladders, toolboxes, compressors and saws. It travelled regularly between Melbourne, Albury and Griffith, New South Wales.

The old ute has given him trouble over the years but never let him down on the side of the road. He rebuilt the motor.

Mark finds it hard to describe why he likes utes so much but says it is just 'the look' and 'the feel' a ute has. In the country he found a ute is particularly good. It is just a case of throwing the gear in, a place to sit in and fall asleep in the back. Very practical. The XD ute is left in the garage a bit more now. He drove it for the first time in a month recently and said it was just great — it brought back so many memories.

He doesn't want to sell it, but says his next-door neighbour wants to buy it so he might, as it means it won't go far and he knows it will be looked after. He really doesn't want to sell it, believe me! But he has to get another vehicle for work and he doesn't want to run three registrations.

Mark's main ute is the dark blue 1985 XF Falcon ute which attracted my eye. He talks of it being 'a Sunday only' ute. He wants to put a V8 in it this year.

He saw it at a mate's place and was impressed. Two months later he got a phone call to say he could buy it. He rushed around and paid an immediate deposit and he's never regretted it.

'I cannot picture myself driving anything else. It'd be a great loss.' He has done nothing to it, apart from maintaining it nicely. He was very worried that some dust might show up in my photos!

Mark plans to put the Falcon under cover and use it only on special occasions, but at the moment he drives it every day. It has 180,000 kilometres on the clock and he doesn't want to add too many more. This is a long-term ute.

I bet that if I go back later this year, he will have another new ute! The blue XF will be in the garage, and he'll be keeping an eye over the side fence to make sure the bloke is looking after the red XD.

The Turnbull family farm near Korong Vale in Central Victoria has about 2000 hectares (5000 acres) of sheep and mixed cropping. It also has five utes. A brand-new Toyota LandCruiser, a HiLux, a VP Commodore and a VS Commodore are the main ones. But when I met Bruce Turnbull, his wife Pam, and sons Paul 21 and Clinton 17, (see *Young Guns*) the first thing Bruce pointed to was the fifth ute.

'Under no circumstances whatever,' they say, 'that ute is not for sale — no amount of money will ever buy it.'

The fifth ute is a little Datsun 1200 that looks as if it should have been put down years ago, but it is a three-generation ute which evokes memories for all the family.

'Pa' Turnbull bought it new in 1975. His wife thought it would be a good, safe and reliable ute. He rolled it on its side on occasion, as the left-hand side of the ute still testifies. But that's another story.

Later it was passed down to Bruce and Pam and it has worked hard over the years. It has carried everything from bags of grain, motorbikes, dogs, sand, sheep and a lot of sheep manure. It still had a lot of manure in the tray when I saw it. It's always been a 'general purpose' ute. Although still registered, it doesn't hit the highways now, except to cross the road to other paddocks.

The speedo stopped when it 'hit the ton' years ago, and it has now done 'at least 400,000 kilometres'.

It has no brakes. The exhaust sounds like a John Deere tractor starting up. It still serves them well around the paddocks, although Pa nearly had a fit when he first saw what Clinton did to it. Clinton is the third generation to get his hands on it, and he decided that he would 'do something after school' with it. Big brother Paul says 'I think he got a bit carried away'.

At the time 16-year-old Clinton took to it with stickers, aerials, lights, home-made bullbar, and huge truck mudflaps. The result is a unique piece of work now known as 'the burn-out king'. It won 'best feral ute' at the Wool Expo last year. His prize was a large esky. The local policeman had an old ute in the competition, but when Clinton pulled up in his, the policeman just shook his head and laughed.

Pa turned up as I was photographing it. A tall erect man, now 86 years old and with

a good handshake still, he is proud of his close family tradition. His own father took up the land here and they have been farming it for well over a hundred years.

He is Robert Gray, his son is Robert Bruce, and eldest grandson is Robert Paul.

Clinton seems pretty happy with 'his' ute. It brings a laugh for everyone and it still fires up okay. The three dogs Sam, Jed and young pup Spot are used to being in the back, although at the time of photographing, they seemed keener on chasing a cat near the house, and had to be called back twice.

'I rolled it on its side once, you know,' says Pa. I never did get to find out how it happened, we were too busy talking and taking photographs. It's a great ute this three-generation mighty little Datsun.

As Pa slid into the old seat behind the steering wheel, he sighs and says, 'Ah, she still remembers me,' as he felt the familiar comfort of the old girl and memories of other times came back.

Dogs, dogs, dogs...

'A ute without a dog is like shagging without a sheila, — kinda lonesome.'

It was too hot and I had a lot of work to do and many miles ahead of me, so I left my two mates Rowdy and Lady at home to look after the missus. I had to be content with meeting other people's dogs. Just as well, 'cause Rowdy wouldn't have put up with another dog within peeing distance of 'his' ute.

Just about everywhere I went ute owners seemed to own at least one dog. In the case of swagman Geoff it was six, Blackie, Daffy, Smokey, Chocko, Woolie and Winger.

In all my travelling around I remember these great dogs and some of their names. I wonder how or why, but each to his own and credit to all — a man and his mate, or mistress and pooch. There was Basil and Rex and Ky. Then there was Rusty and Red, Skipper, Nipper, Snifter and Ted. I met Jeff, Fred, Dog, Buster and Blue.

The tyres of my ute have been peed on by every imaginable sort of dog from Broken Hill to Bendigo, Mildura to Melbourne, and beyond. Whenever I come home, my two dogs sniff me out wondering at the doggie smells and where the hell I've been.

Believe me, dogs know if you're a dog lover or not. Most check you out discreetly, some are all bark and bluff, not bite, others wary and shy. Some are over the moon to see you and will play, lick and wag nonstop. Many are sad to see you go. They know when you're leaving and show a look of 'come again'.

I hadn't had any bites — until the second last day. A blue heeler named Sally didn't like the look of my camera. I was on her turf and near her ute! She gave me a nip on the hand. It had to happen but we parted friends.

Dogs just love to please. They accept you for who you are. You can grumble, yell, annoy or ignore, or come home in a bad mood but they have an inbuilt instinct to please you. They love to show you how thrilled they are to see you. They get excited; pups will even piddle with glee. We can learn a lot from dogs.

They can work all day for you. They love to help round up sheep or cattle and show you how clever they are. All they want in return is a good bit of tucker, plenty to drink, a pat and a kind word, 'well done'. They will fight to the death in order to protect what's yours, because what's yours is theirs. For life.

So they jump on your best clothes with muddy paws and they nip your ankles just to say 'G'day'. They bark and bark to say 'Hey, it's me!' And when you go, they howl in despair, pleading 'don't go'. Their bodies can't keep up with their wagging tails when you come home, and slobbering is just a happy laughing sign of welcome.

You can tickle their tummies and they'll give you a smooch. Throw a ball or stick, they'll retrieve it or run off in fun. They'll snuggle in on a cold winter's night and snore or fart or moan with delight; sound asleep but up with a start at the slightest noise. They'll stretch and yawn, snarl or wimper. Man's best friend and woman's delight.

The very least you can do, is to tie them on safe when you're driving your ute.

'FJ'

'Utes are like your best mate — you never part.'

It's probably just as well that 16-year-old Bianca Siemering crashed her father's 1990 Holden Rodeo into the garage. 'If it had been the other ute, she wouldn't be sitting there smiling now,' mocks her father, Brian.

The 'other' ute is the family's pride and joy. A 1956 FJ Holden ute.

Bianca's story is that she was just practising her driving and thought she'd put it away, 'but the shed jumped out in front of it.' She smashed the mudguard, bumper, and the garage door could no longer be closed. 'It could have been a lot worse though. She was headed straight for the FJ ute, but hit the garage instead,' says her mother, Trish.

Brian is a self-employed builder, and some 16 years ago spotted the FJ in an open shed in Wycheproof, Victoria, when he was working there building the high school. The ute was in original condition, with only 86,000 miles on the speedo. The owners had bought a new ute, so had just parked the FJ in the shed and let the registration run out six months before Brian bought it for $1000. It started first go. He took it home to Castlemaine but it didn't stay there long.

A good mate, Eric Peddler, said he'd restore it, which suited Brian. He always knew he'd get it back from his mate so the FJ moved to Ocean Grove on the south coast. Eric got stuck into it and spent a lot of money restoring it and drove it around town for the next 15 years!

Brian didn't have the money to do the ute up at the time but, when his mate decided to buy a block of land years later, Brian handed over $8000 towards the land, and took the ute back. That's what mates are for — to help each other. It was a special occasion and there are photos of Eric handing the keys of the ute back to his mate Brian. I dare say a few light ales assisted in the celebration.

'There was never any question we'd get it back,' says Brian. 'Eric still comes up three or four times a year to drive it. Even if I

wasn't here, he'd just take it out of the garage.'

Eric's wife, Margaret said, 'if you sell it, I'll shoot you.' She needn't worry, this ute will never be sold. Brian loves it, Trish loves it, the kids love it, and Deefa loves it. She's the great 12-year-old long-haired shepherd. The name is a typical beaut Aussie name. 'Deefa' — D for dog. Get it?

This ute moves! It runs on a completed speedway race motor. 'A 202 fully race balanced, jetted, with extractors, and a Hadfield conversion 5-speed Celica box — it'll run off the clock — 120 mph-plus, and it sits on the road real well, like a tonne of bricks,' says Brian.

The FJ is painted a BMW gun-metal silver-grey, with mag wheels, top interior. There's nothing but admiration all round the table as we talk of the old girl. You can understand how much Brian likes the FJ when you hear him describe the other vehicles in the garage. 'The Rodeo means absolutely nothing to me. The Commodore is a good car, but it's just shit compared to the ute.'

Says Trish, 'It's got a bit of charisma.'

And what of the other vehicles in his life? He is a real Holden man from way back. He is quick to point out that, 'I've never been without a ute or a motorbike. My first was a Holden FX, an HT, two HRs, an HJ, HX,

WB, WB one-tonner, a WB one-ton tipper, and one mistake was an XF Falcon — the worst-ever piece of crap, the doors used to fall off, it was a bucket of shit.'

Trish drives a VR Commodore to the local hospital where she works as a nurse, while Brian uses the 1990 Rodeo. There's laughter all round when the conversation comes to the Rodeo ute. It's 'just a work ute and is never looked after'. You can hardly get

inside it for junk. Brian often hears comments like 'so you're still living in it then?' Bianca says that 'even mail for the last 12 months gets lost in it'.

Trish describes Brian as 'a boy and his toy' but she also says 'Utes are the best'.

As our conversation winds up, and we finish our cuppa tea, it is 11-year-old Rachel who sums up the family's feeling about the ute best: 'It's good, and it goes fast'.

FRANK THE RABBITER

I met old Frank only once. He was stopped in a battered old 4-wheel drive and I pulled up to offer help. It was on the hard dry claypan and black soil country in outback New South Wales. It turned out he had just stopped in the middle of the road for a break.

Mine was the only car he had seen all day. I was doing historical research for the Department of Conservation, Forests & Lands. He noticed the government number plate and said dryly, 'G'day, long way from the city, aren't you?' He was right, I was, and heading much further into crow and dust country along some of the loneliest driest tracks known. But I had finished work and was on my way to a nearby sheep station for the night. Typical of outback hospitality, he offered to share a cup of tea.

When we parted nearly three hours later, I felt a happy man, having shared in his memories. He told me yarns of his long and varied life, rich and poor years. To me, he was a rich man — he had lived life to its

absolute. Now retired, he spends his days collecting old bottles and fishing in the Darling.

Amongst his many yarns was the one about his Dodge. He showed me photos from an old envelope of his once-beloved 1924 Dodge Tourer, which had been unceremoniously cut down and home-made into a ute.

'By cripes, I had some good times in that old bus,' Frank said as he slopped his tea down his chin onto his collarless shirt.

'I remember driving through Balranald on a stinking hot day with a big load of rabbits on the back that I had shot. Blood was dripping through cracks in the tray. The frigging noise coming from the rear axle was so loud everyone was looking to see what the commotion was. I pulled me hat lower over me eyes and prayed I could get through town quickly, before the coppers got me.

'As the ute got to an intersection it grunted, farted, coughed and stopped dead. I nearly died. Blood still dripping, and smoke from the hot motor everywhere.' Frank roared with laughter, nearly spilling his tea again.

'I got going in a cloud of smoke. Some young kids were on the footpath and one yells, **"Hey Mister, ya' back wheel's going round and round."**

'He thought he was being smart in front of his mates. Without thinking I yelled back "So would you, with an axle shoved up your arse, Sonny."'

Gary 'Fuzz' Conway is a country boy. He bought his first ute when he was 16, while still at high school. It was a $100 light blue 1960 FB Holden with 66,000 miles on the speedo. It had no brakes, so when his father towed him home with a V8 HZ, there was a fair chance something was going to happen. His father's advice was 'don't worry about the brakes, just use your clutch'.

'Dad tears off with me being towed. There was no horn in the FB and, needless to say, when he finally stops, he sees I'm so close to him I'm nearly hitting his vehicle. He roars the crap out of me for nearly running into him!'

Fuzz wanted to show it to his 15-year-old schoolgirl sweetheart, Maree, but she said she didn't want to see it 'until it's done'. It took Fuzz two years and $2000 to restore the FB. And Maree's verdict? 'I thought it was fantastic!'

Fuzz learnt to drive when he was 10; Maree when she was 11. The childhood sweetheart is now the mother of their four children, Courtney 10, Alana 8, Alexandra 5 and Timothy 4.

'We went everywhere in the FB, all our friends did. None of the others owned a car, it was the vehicle for all of us,' she says.

'I had five sheilas and myself in it one night,' laughs Fuzz. 'It got pranged and I had run out of petrol. My apprentice, who had had his licence for only three days, was towing me back to the workshop. We pulled up at some lights. I got out because I had driven over the tow rope and it had caught under the wheels.

'Next thing I know he takes off, with me on the outside. I'm chasing and yelling for him to stop. He had the windows up and didn't hear me. I couldn't get back into the ute. It finally pulled to the left, the rope broke and the ute ploughed into the stop-lights while he kept going. It was stuffed.'

About 12 months ago, the ute was sold for $200. It had 160,000 miles on it.

There are now two utes in the family and a stationwagon for Maree, the kids and of course, we can't forget Frank, the placid 14-month-old german shepherd.

Fuzz is well able to fix up utes. He operates his own business as a qualified mechanic, doing LPG conversions, mechanical repairs and roadworthy tests. He and two workmates carry out repairs on about 75 cars a week for nine local caryards, as well as their own customers.

One of the two utes is a 1978 HZ Holden 'with not a straight panel in it'.

'I bought it up the Mallee. Originally it was bought new by W. W. Watson of Berriwillock. Then "Hoss" Ballentine in Hopetoun had it. He traded it in to the original dealer, who then sold it to Harry Ackland of Woomelang. When he died, his son Geoff, got it. My brother Tim married Geoff's daughter Zoe. I bought it from them in Rutherglen in 1996 for $2000.

'When I got it, it was stuffed. We got a wheelbarrow full of mallee dust out of it. Everything was stuffed. It took 18 months to restore. The body still needs to be done. It had 120,000 kilometres on it. It's now got a fully reco motor and done another 8000 kilometres.'

The other ute is a 1962 EK Holden (see page 59) yet to be restored. Fuzz paid $900 for the ute which is still only undercoated. Fuzz has got spares, but no motor, gearbox or seats. He plans to restore the EK to original condition and has plans that the final result will be a standard job. Maree says

she'd 'rather he had fixed up the FB', and she prefers utes 'a bit more lairy, but he wants it an original colour — white.'

It is obvious that Maree is a ute girl. She did her training as a Registered Nursing Sister at Mildura Base Hospital, working in the intensive care unit, and she has fond memories of all the utes they've had, including a WB Holden which used to let so much dust in, it covered the baby pram.

'Mum's got a Commodore ute on the farm, and Dad has a Mazda ute. I don't know why but you always feel you can drive faster in a ute.'

Jim Haines paid £1175 for a brand-new brown Dodge ute in November 1950. Jim's fiancée Grace asked him, 'How'd you come by that?' Mrs Haines smiled, 'But he never really did tell me.' It was the start of a love affair that lasted many years. Jim and Grace were married in August 1951. Their ute took them on a three-week honeymoon to Castlemaine, Ballarat, the Grampians, Mt Gambier, Portland, Framlingham, Great Ocean Road, Melbourne and back home to northern Victoria.

They moved onto their farm where Mrs Haines still lives and share-farms with her family; Jim died in 1991.

Their ute was the only vehicle they owned for a number of years both as a farm ute and as an off farm vehicle. On a holiday to Adelaide, Victor Harbour and beyond Jim added a sunvisor and he had an aluminium canopy made to fit the back of the vehicle when they went on camping holidays to Eildon, Mt Buffalo, West Wyalong, Hillston, Leitchville, Albury, Henty, and to her hometown of Sea Lake with a ten-week-old baby in a cane dress-basket. The ute was a faithful companion for many years.

Although the registration lapsed in 1972, the ute was still used as a workhorse on the farm until 1990. Their son Allan, and son-in-law Russell, put a new motor in the Dodge in 1980 as a birthday present for Jim. Russell remembers it took them two

days to fit. 'It was a shocker, such a big motor to fit in such a little hole.'

The tailgate was removed in 1972. The gear shift, although very sloppy, continued to serve its purpose. In later years the ute needed to be handcranked and there was a ritual to get it started. You had to pour a

small amount of petrol into the carburettor, and with choke partially out it would start every time.

'Jim always believed you should drive on the gears not the brakes.

'I think towards the end it became a bit tired, Jim got tired of keeping the maintenance up.' By then it had brakes on only two wheels — it was impossible to get the rear hubs off, and parts were hard to get. They decided it was time.

In 1972 a 12-month-old XY Falcon ute was purchased. It remains a top workhorse on the farm. When I visited Mrs Haines it was in Bendigo being serviced by Allan, who is a mechanic. The Dodge was much bigger and could take hay and grain for feeding sheep in just one load but the XY ute can't carry both at the same time.

Under the cobwebs that cover the Dodge dashboard the speedo reads 82,766 miles, 'That's the second or possibly third time around the clock,' laughs Russell.

The ute has seen better days but Mrs Haines agreed that there were many good memories for her and her honeymoon ute. Mrs Haines looks quietly over the remains of the Dodge, standing beside it proudly to have her photo taken. We leave the ute to its resting place and head back to the homestead for a nice lunch. We have gone back in time and have rekindled some memories; she has a look of happiness in her eyes and warmth in her voice.

'The kids all learnt to drive in that old ute.'

In Search of...

It is true. Ute owners are a breed all of their own. I was munching on a Chiko roll in Benalla, being earbashed by a milkbar proprietor who knew of the 'ute-of-all-utes'. Twenty minutes later I managed to move backwards out of his front door as he started on yet another ute story. He had written instructions for me to find this ute-of-all-utes on a Four n' Twenty pie bag.

I headed off with great expectations following the instructions miles and miles down a dirt road. I wasn't lost but mighty confused for a considerable time before I found the front gate of what looked like a deserted farm.

An incredibly ferocious dog prevented me from getting out. With its front paws against the door, all I could see through the slobber that the flea-bitten mongrel had deposited on my window were great fangs. There was no way he was going to let me out; the mongrel had death — and me — on its mind.

I blew the car horn for about five minutes but no one came from behind the ancient cypress hedge that surrounded the homestead. Eventually I spotted a decrepit Holden ute in a shed parked next to an International combine harvester, and this fitted the description the milkbar owner gave. It was certainly not the ute-of-all-utes. I left in a cloud of dust, abandoning the mutt to his domain and, I hoped, coughing dust.

I would be sent on many other red herring chases. Everyone seems to have a ute story. I was often late arriving at my destination because I was always being side-tracked.

Only three people said 'no' to being included in this book: two middle-aged codgers and one arrogant young buck. Most people were very obliging. It never ceased to amaze me how enthusiastic people were; a few would have paid me heaps just so that they could be in this book (they didn't, and they are!).

One thing I probably enjoyed too much was the look on people's faces when I pulled up beside them on the open road and motioned for them to pull over. Most had guilty looks even when they'd done nothing! One bloke said 'Christ mate, I thought you were the bloody fuzz, you nearly gave me a bloody heart attack!' My response on numerous occasions — and one which always brought a laugh — was, **'When's the last time you saw a copper with a beard, long hair and driving a ute, mate!'**

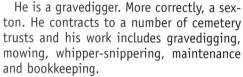

Alan Graham was riding a motorbike through the countryside in May 1979. He stopped and asked a bloke if he knew of any available work.

'Yeah, you can have my job,' was the answer. What a surprise!

Alan has shown his reliability; 19 years later he still holds that job.

He is a gravedigger. More correctly, a sexton. He contracts to a number of cemetery trusts and his work includes gravedigging, mowing, whipper-snippering, maintenance and bookkeeping.

His workhorse is his trusty yellow 1984 Toyota HiLux, loaded down with an assortment of tools — crowbars, shovels, fuel, whipper-snipper.

He has had the ute since new in 1984 and has done nearly 250,000 kilometres. It is to be 'retired' from fulltime work as he hopes to get a new twin-cab ute to accommodate his growing family. He'll keep the old ute though.

The HiLux is a 'hundred per cent work ute' which is essential to his job and, importantly, 'it is small enough to manoeuvre between the headstones.'

A stock crate for the back tray makes the ute very handy if you live on a few acres, as Alan does, to carry fencing materials and the like. He has four goats and two pigs and the ute is also 'good for getting lambs from the saleyards'.

After not seeing Alan since I first interviewed him I rang to get the latest news and heard that he has gone on to buy his HiLux twin-cab, with CB radio to keep in contact with his wife Anne. The original ute is still a part-time worker.

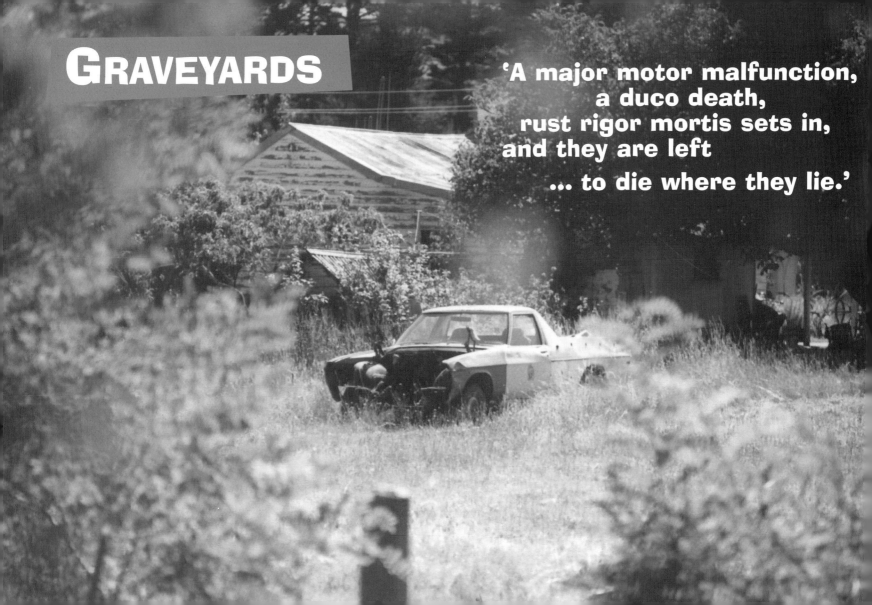

GRAVEYARDS

'A major motor malfunction,
a duco death,
rust rigor mortis sets in,
and they are left

... to die where they lie.'

GREAT DIVIDE UTE

I'd passed the farm many times. You could tell it was an old family property. It is situated in the high part of the great dividing range, and will soon be even more isolated as a new road will by-pass the area altogether.

I'd also seen the grey HD Holden with the canopy on the road many times. One day I decided I'd call in. I'm glad I did. I met Peter, a bachelor in his early 60s. He's the third generation of his family to own the property and he now lives on the remaining 33 hectares (81 acres).

Peter loves all animals. Kangaroos come up to feed each night. His cows and even the three bulls are very friendly. Cats and chooks are many, and he recently saved a family pet from certain death, reuniting it with its family. His love of animals was once written up in the local paper.

I found Peter and his dog Bristow easy to like. Peter's been having a bit of a problem with his heart of late, so I moved a heap of things for him. He was very pleased.

He knew I was interested in history so we walked around the property, followed by Bristow, three bulls and a number of cows and calves. We looked at the remains of his first family home which was built in 1856. We wandered past an assortment of cars, utes, tractors, and all sorts of farm machinery. He's selling most of it now and the rest will be gone within a few weeks. The family farm is up for sale but he hopes to remain on the original homestead site. He proudly told me that if he dies tomorrow

the farm is willed to the state to become an animal refuge.

As we walked there seemed to be more and more vehicles popping up from behind trees — an FX Holden (totally stuffed, but I bought the grille as a momento), an XT Falcon, a 46 Chev ute, an FC Holden wagon, a Mark II Zephyr ute (rear only), and more. Soon they'll all be gone. He rattles off some of the vehicles he's owned — a 35 Chev coupé, a 51 Plymouth, a number of 46 Chev sedans, and others. He showed me photos of them all.

In the big shed was a surprise — a 1939 Chevrolet ute, which his father had bought for £575. It's now been sold with all spare parts. The 39 Chev has worked all its life. The motor is original, like everything else. Even the original manual is in the glovebox.

Over the years it has been used to carry all sorts of stuff including 30 bales of grass hay at a time and many thousands of tonnes of firewood from the forest. His father was a forestry officer for many years.

He says the ute used to be so heavily loaded it broke axles more than once.

A lifetime of hard work hasn't stopped the old girl. Put a battery on it, and she still fires up. They certainly knew how to build 'em in those days.

Rusting away under a tree and heaps of rubbish is the remains of a 1946 Chev ute, just like the one my father owned in our family grocer shop. Peter bought it for spares and it's been ripped off ever since. Its days are numbered.

One vehicle that won't be sold is Peter's HD Holden ute with canopy. He's had it since new, and it is now home for his best mate Bristow.

'I'd be buggered without that ute — like a bloke without a shed. No bloody good. It's

my life, it's part of my life,' he says passionately. 'And Bristow lives in it, that's his home, he'd do no good without that ute. Bloody best mate you can have. He used to belong to a chef who had to move to Townsville, so I said I'd take him. He's eight now. He's a great guard dog. If he meets a no-hoper he very quickly latches onto their arm and won't let go until I tell him to.' Peter laughs. Bristow had been sniffing me and was now allowing me to pat him: I was relieved I wasn't in the no-hoper category.

Peter was a motor mechanic for 30 years, as well as working the farm. He loves all sorts of vehicles. Soon he will have only the HD and a Falcon. The HD was bought new in the 1960s; it now has 100,263 miles on it. It has original motor, clutch, brake liners, and is still going strong!

He put the canopy on which he bought from a bootmaker ('he's well dead now') for $20 about ten years ago. It's off a Valiant but he adapted it to fit the Holden. The sliding glass windows which keep Bristow safe and out of the weather are from the old Mark II Zephyr ute.

The following week I returned to see how Peter, his dog, bulls and cows had fared in a major bushfire. His family have been there since the 1850s and fire has always been a threat. He had survived yet again.

GRUNTER HUNTER

I spotted Luke's ute in a carpark and left my business card under his windscreen wipers with a note. I caught up with him a few weeks later, on a dusty track in the bush behind a golf course. He had his girl-friend, Amber, with him as well as his two dogs, Wolfe, the six-year-old wolf-hound–bull mastiff cross, and Skitch, the 18-month-old great dane–bull mastiff cross.

Now, I've got to tell you, I knew these dogs were pig hunters but I didn't know what to expect. They looked as if they could kill. Luke let them out of their cage on the ute and they both headed for me. **If you've never had your crotch sniffed by two bloody huge pig killers, you don't know what you're missing!**

After we sussed each other out, we made friends. I got heaps of big slobbers all over my face from the two big sooks. They were great, and tails wagging they thought I was great (I hope).

Luke bought the 1980 Toyota HiLux ute secondhand four years ago for $5000, but it's cost him about $11,000 all up. He says it was a pretty ordinary old farm ute when he got it, but it is okay now. He's gone through three 4-cylinder motors and now runs a Holden red 308 V8 motor in it, with turbo 400 auto.

'It's got a Ford 9-inch diff with LSD centre, high ratio diffs, a power winch off

a LandCruiser, B. F. Goodrich 33 x 12½ inch ideal tyres, and a number of other modifications. Home-made spring hangers and WB front coil springs helped to lift the back up about 4 inches.'

Luke did most of the work himself. He is a fourth-year aircraft mechanic for Ansett, working on aircraft engines such as Boeing 737, 767, British Aerospace BAE-146 and Airbus A320, so working on his own vehicle comes easy.

He also built the rear tray and dog cages. The roll bars he had made. 'A month after I had them installed, I swerved to miss a kangaroo and rolled the ute. Without the roll bars the whole cabin would have been destroyed.'

When Luke was younger he hunted with a bow and arrow and a trail bike in state forests. He has hunted pigs for the past three years. At least twice a year you can find him in outback New South Wales and across the border into Queensland.

Amber's family live on a sheep farm at the edge of a mountain and forest, so she too is now learning about pig shooting. She showed me her bruised shoulder caused by

using a shotgun.

Rough country life and travel to places such as Wilcannia, Broken Hill, Booligal, Back o' Bourke and bey-ond have caused Luke many prob-lems. Broken axles, snapped bolts, dropped tail shafts, blown head gaskets, stuffed wheel bearings and many other breakdowns are all part of the excitement for the 'Grunter Hunter'.

'One drama that didn't impress me hap-pened one Thursday. I'd paid $1300 for wheels and tyres, and the next day, Friday, en route for a shoot, I staked a tyre even before I got to where I was headed.

'I'd never buy another HiLux; the next ute will be a Toyota LandCruiser, but I will build the same modifications for shooting, though — power winch, big spotlight, roll bars and dog cage.'

When Amber drives the ute, she always gets a lot of looks, 'I just think it looks good — really great. It's hard to start, hard to drive and hard to stop. But it's a power trip for a girl to drive. Plenty of looks. **I pulled up in the street one day and this girl says to me, "so you're the Grunter babe, are you?" I just laughed.'**

GULGONG JIMMY

The township of Gulgong, New South Wales should declare Jimmy Grimshaw a 'living treasure.' At 77 years of age Jimmy has rebuilt his life after a massive stroke left him unable to walk or write. Now a few years later, you can't stop him. With the aid of a cane he keeps busy, as many people in town can testify.

Many homes have his beautifully crafted furniture. He has made over thirty grandfather clocks. Self-taught, he is a real craftsman. He has always loved wood, cedar in particular. He spends his days restoring all sorts of timber, from tables and chairs, to countless other bits of furniture. He said his work has kept his mind active and helped him survive and recover from his stroke. He should have a book written about his life.

A wool classer for 50 years, he still lives in the home where he was born. He never married, but has many friends all over the country. A real gentleman, he is the kind of bloke they will talk about with affection long after he's gone.

In his well-equipped sheds are some interesting vehicles that he has beautifully restored with a mate. There's a 1939 2-door Chev slope coupé (that he bought from a church minister — one of only eight in Australia); a 1948 Vauxhall sedan (owned by an old lady since new), his Suzuki Sierra 'knockabout' ute, a Magna stationwagon, and the one he's promised to sell me someday for the *Beaut UTES* Museum, is his restored 1937 Chev commercial dropside ute. All vehicles will go to good homes eventually.

The Chev ute he bought in 1949; it has 'been everywhere'. Jimmy used it on his farm for many years and it was a real worker. It was his only vehicle and he used it relentlessly. When he retired he decided to restore it and it is now like new and Jimmy is very proud of it.

Jimmy is in safe hands. People in town keep an eye on him, friends turned up to see who I was and his neighbours never miss a trick. I heard about him from a lady in town who has some furniture that Jimmy restored. Everyone likes Jimmy. My wife and I are going back to see him in a few months to share a few more of his wonderful yarns and another pot of tea and biscuits. She'll love him too and I think we'll probably find him working down in the shed on another project. And keeping 'my' ute clean.

HALF MILLION MILE MAILMAN

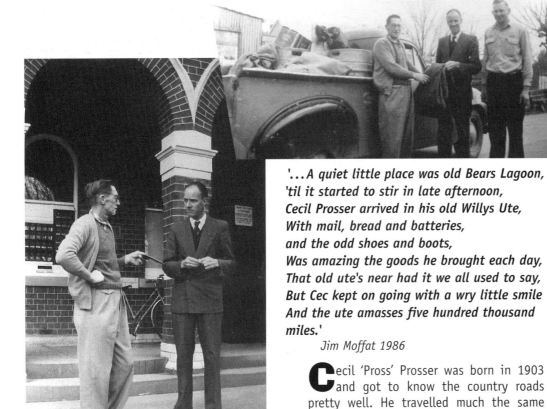

'...*A quiet little place was old Bears Lagoon,*
'til it started to stir in late afternoon,
Cecil Prosser arrived in his old Willys Ute,
With mail, bread and batteries,
and the odd shoes and boots,
Was amazing the goods he brought each day,
That old ute's near had it we all used to say,
But Cec kept on going with a wry little smile
And the ute amasses five hundred thousand
miles.'

    Jim Moffat 1986

Cecil 'Pross' Prosser was born in 1903 and got to know the country roads pretty well. He travelled much the same mail route for 32 years and half a million miles before his contract expired in 1959. His first mail contract for the PMG (Post-Master General) began in 1926, and his route was from the Jarklin Post Office to Inglewood, taking in the small communities of Bears Lagoon, Serpentine, Salisbury West and then reloading at Inglewood for the return trip.

In later years he moved to Inglewood so the contract was reversed. When he went out in the afternoon, as well as the mail, he'd carry bread, medicines, groceries, vegetables, beer in barrels and many other goods. On the return trip the mail would be surrounded by cream and eggs to be delivered to the railway station for transportation to Bendigo.

He worked six days a week and often gave schoolkids a lift to school. In the early days most of the roads were dirt, which quickly turned to bog in the wet. It was nothing to be bogged a couple of times a day in winter. In all his 32 years of service, he failed only twice to get the mail through due to extensive flooding.

His first ute was a T–model Ford, followed by a 1926 Chev, a 29 Chev, a 37 Willys, and finally an Austin A70. When his mail contracting expired, he took on school bus driving until he retired. He died in 1970.

HARD-ROCK MINER
& THE OUTBACK
BRUMBY

I met Allan McLennan leaning against the bar of the Silverton Hotel in western New South Wales the day before an Australia Day holiday. We shared a bit of a yak with Ines the licensee, whilst having a quiet ale. It was sunny and blowing red dust outside.

Allan was born in Wilcannia in 1941 and he's been around a bit. He's worked as a shire worker, a stationhand, publican and hard-rock miner for Broken Hill North until he was retrenched after 29 years. He was one of 500 to lose his job.

'Anyone who knows mining knows that coal mining is nothing like being a hard-rock miner.'

Allan and his wife Sue were in the midst of restoring an old stone and brick house in Broken Hill when he lost his job. They both became addicted to restoration and won the town 'Best Restored Home' a few years ago.

He invited me to see his second home, an 1898 miner's cottage in Silverton, which he and Sue have spent the last two years restoring and setting up as tea rooms and a gallery. Corrugated iron, local rock and sand have all gone into making an unusual and attractive cottage. Allan's building practices are unique: walls of mullock, granite and local rough stone.

Due to Sue's ill-health they have to sell the cottage and will spend six months of the year in the Broken Hill home, and the cold winters somewhere on the coast.

Like most ute owners Allan has a mate in his ute, Basil the little wirehaired terrier, a friendly lovable pooch.

Allan's ute is a Subaru Brumby which he bought brand-new in 1990. I commented on how I didn't really like Brumby utes much; they just don't seem to be designed right but Allan was quick to point out that the ute has been used to carry rock, granite, and sand from around Silverton and Broken Hill, and heavy loads have never been a problem. Try to push a heavily loaded trailer uphill and see how your ute goes. Easy to pull a heavy trailer, but try to push one up (backing I presume). **He slips the ute into 4-wheel drive and it has moved mountains!** Never lets him down.

Allan was loading sand from a creekbed

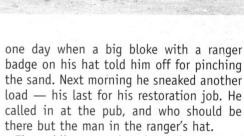

one day when a big bloke with a ranger badge on his hat told him off for pinching the sand. Next morning he sneaked another load — his last for his restoration job. He called in at the pub, and who should be there but the man in the ranger's hat.

The publican mentioned to Allan that he had seen him hard at it loading sand that very morning by the creek. Allan tried to quieten the publican down, changing the subject. He pulled his hat down over his eyes, and quickly left after a beer. Later he asked the publican who the bloke was and was told it was Chief Detective So-and-so! Back then the police were also honorary rangers. Allan nearly died. It was enough to end Allan's sand-mining days!

Ged O'Connor stands a lofty two metres (6' 8") and is all soft and squishy (well according to his wife Gai anyway). I spotted his pale blue 1966 HR Holden ute parked in Cowra, New South Wales and saw Ged go into a milkbar, so I did a U-turn and pulled in.

I nailed him as he left and we had a chat on the footpath outside the milkbar whilst my steak sandwich was being made. It was dark so I made an 8.30 a.m. appointment for the next day to call on him.

Ged and Gai welcomed me as we did a photo-shoot, then with a cuppa tea in hand, we all sat down to hear of the ute.

Ged lived on the 810 hectare (2000 acre) family farm until he sold recently. He bought the ute four years ago for $500 from a lady in Canowindra. It had sat in a shed for 13 years. He loves it.

It has front end disc brakes, a 186 motor, a great comfortable seat from a HG Premier, and new diff centre. Ged sold his motorbike so he could buy it. Just after he paid for it, a mate told him of another white HR ute that still had registration on it for $600, so he bought it too.

'One night I was in the white ute, pissed as a fart after 'a fair few beers' I was cutting across the paddock heading for home, I saw a dark shadow on the ground, but it was actually a dam full of water — and I drove straight into it. The water covered the motor and there I was sitting in water up above the seat with more water rushing in. I had to climb out the window.

'The ute sat there for a while before I finally pulled it out, drained the motor, towed it around with the tractor for half an hour. The motor freed and after another 45 minutes I managed to get the motor to fire. I then drove it for another three years. It wore six radiators out. It had a bit of a hard life. The engine mounts rusted out. Once I had a pushing match with a 4-wheel drive. The bonnet was bent up, and I ended up having the radiator tied on with rope. It's still going on a mate's property. I sold it to him with other Holden parts for $30.

When Gai saw the pale blue HR she wasn't impressed. 'I absolutely hated it when he first took me out in it. Old, uncomfortable, no six-stack CD player. Only an old valve radio. No heater, no air-conditioner, no power steering. Too hard to drive. Then Ged put the new comfy seat in. That did me in. I love it now. Ged wants an FJ Holden now. I don't want him to sell this — but it's one or the other.'

Gai too, is into motorbikes. Her first bike was a Suzuki RM80 when she was 12 years old. In fact, she's had over 20 motorbikes in the last 23 years. She taught rider training in Sydney where she was the only female instructor. When she'd take her helmet off she used to get comments like 'God, — it's a girl!' or 'What's a little girl like you riding a big bike like that!'

She has a message to the girls out there who get a hard time from the fellas: 'Stick ya' middle finger up, hold your head high, and f....... enjoy yourself!'

Ged and Gai met at a motorbike rally. She first spotted this tall bloke leaning on the bar, and thought no way she was going to go near him. But they were teamed up as pool partners, and the rest is history. 'He's as soft as they come, he even wrote me a poem.'

Ged is equally impressed with his true love. When I asked how long they'd been together he replied 'two years and 13 days.'

In the backyard I met playful 'Cisco', the two-year-old blue heeler cross. Out of the garage came Gai's car and the mighty Harleys. Ged has just bought a brand-new Harley Davidson 'Fat Boy' 95th Anniversary model, and Gai's is a Harley Davidson Sportster 883 model.

As soon as Gai finishes her nursing studies they will be gone from Cowra. They've bought a large house on 30 hectares in the northern sun.

'JB' & THE RED HILLBILLY

Justin Brown is not your ordinary businessman. He left school at year 9 (form three). With little education but lots of optimism he set out to work. Now, at 32, single and owner of two successful businesses, 'JB' knows what he wants out of life.

He has operated a tree and stump removal business in Melbourne for the past ten years, and in the last 12 months also started Mini Ag Services specialising in any jobs that require a front-end loader, trench digger, ground ripper, mower, post-hole digger or super spreader — you name it, he can do it. He operates this business in Central Victoria, where he is building his new home on 10 hectares (25 acres).

He works hard because, as he put it, 'I like money so I can do the things I want.' He has a strong commitment to planning and taking 'one step at a time'. His various pieces of machinery have to 'pay for themselves'.

JB is mad on the 'old bush Australian' lifestyle and collects blacksmith tools, old hats, old timber, and other Australiana. He loves his Ford F100 ute and his three dogs Digger, Jessie and Redgum.

Digger gave me a wide berth; he's a one-man dog and let me know it in no uncertain terms. Jessie is the most intelligent of the three dogs. All go with JB wherever he goes, every day, seven days a week.

JB restores 4-wheel-drive vehicles for re-sale and had one in the shed under restoration when I visited him on his property in the hills overlooking a pretty valley north of Melbourne.

Justin says his bright red F100 ute is a 'special built' rig. The bullbar is a one-off and one of the meanest you'll come across, as a number of small vehicles that have tried to leave their mark on it can testify. The police have tried to put it off the road, but he passed the roadworthy test. Police don't like bullbars or big lights.

It has a big 351 Cleveland motor, air-conditioner, airhorns, air-operated rear lift suspension which runs off an air tank and compressor, and a purpose-built rear tray and cabin for the dogs.

People who have never met him know his ute by the red and white number plate HILBLY.

JOCK'S MERCEDES

Jock's place is easy to find. It's a couple of hours over the New South Wales–Victoria border past the end of the bitumen and across the saltbush plains. Old telegraph poles dance in the heat haze on the horizon. You know you're on the right track when the rearview mirror reflects only dust. Turn left at the crossroads and head to where the sun sinks. You can't miss the place (it's the only place in sight).

Jock left his parents' property when he was 18, and bought his own sheep station next-door. Ten years later he bought another next-door as well. Now, six kids later, Jock is well entrenched in the district. He also bought a house on a bit of land a few hundred kilometres down the track beside the river to relax and have somewhere for the kids to swim.

Jock and Kerry own about 40,000 hectares (100,000 acres), which was once part of one of Ben Chaffey's properties (see *T-Model Jackaroos*). They'll need more land soon — with daughter Jane 21, and five sons Dallas 19, Leon 18, Rhys 14, Ryan 12 and Zack 10, there's a fair chance there's going to be another generation to carry on.

I first met Jock and Kerry back in the 1980s when I set up a museum on the property whilst doing historical research. It took me two trips and two weeks sifting

through papers in dozens of old boxes covered in dust and bird poop in numerous sheds. The history of the property goes back to the mid–1800s.

On this third trip — after ten years of not being out here — a hell of a lot has happened. More sheds, bigger house, more vehicles — more utes. There's a 46 Dodge truck, a 20 McCormick Deering tractor, a 46 road grader, a bulldozer, a front-end loader, at least six motorbikes, a dune buggy, a Suzuki Sierra ute, a Suzuki SJ40 full-cab and 'Lumpy' the Hillman (that raced at Calder Raceway years ago). In another shed is a Bunyip ultralight aeroplane. And that's just the start!

We moved from one shed to another, utes moving here and there, kids fighting over who was going to drive, sheep dogs, Lassie,

and Dusty, excitedly a part of the action, as was Spot the little house dog. The other dogs Whisky and Red were not impressed with being kept tied up.

Jock was always a Holden man. He's had an HD, HT, HG, HJ and Commodore. There's also been two Mazda 1000s, a Datsun 1000, a Datsun 1200 and eight Suzukis.

'Jock never sells anything,' laughs Kerry. 'You're just a car-a-holic'.

Then Jock 'went Toyota'. He's had a couple of LandCruiser utes and two Sahara wagons. Now parked under shelter next to the house are the current Toyota workhorses, a LandCruiser ute, a twin-cab HiLux and a Sahara wagon. There's also an old Corolla 'paddock basher' which belongs to Rhys.

What I came here especially to see again is locked away in the shed right next to the house. It is a special ute. It is one of only four in Australia. A 1952–55 Model 170SD diesel Mercedes ute. This is no home-made ute; this is the real McCoy.

Originally imported into Australia by a station owner who used it as a runabout for a number of years, Jock's father bought it at a clearing sale in 1961 for £150, the last year it was registered.

Jock's father drove it for about 1000 miles, but did a universal. 'Fancy putting bloody rubber universals in a vehicle' was his comment as it was shoved in a shed.

In 1966 Jock acquired it from his father and rebuilt the motor and all mechanicals. At a time when he was earning about £10 a week, Jock imported four pistons for £80. He drove it for about 200 miles, and for the last 30-odd years it has stayed in the shed.

The ute needs a bit of paint and body work. Everything is there, a hundred per cent complete. It will be fully restored to mint condition says Jock, as a retirement project.

A lot of people have tried to buy it but no amount will ever buy it. Everyone in the family loves it. Jock's Dad was wondering if I was going to buy it when Jock told him I was coming up to see it again. But no, it will never be sold.

Let's hope young Zack, 10, doesn't get his hands on it too soon. He told me he is going to use it as a 'paddock basher'! He learnt to drive when he was 5 years old, says Kerry. And as I prepared to leave, Zack gunned his motorbike and disappeared in a cloud of red dust across the yard.

My time was up. I promised that my wife Jan and I would return. Five dogs barked their farewell. Now it was my turn to create some red dust as I headed off into the kangaroo-infested twilight towards home many hundreds of kilometres to the south.

Most ute owners do not wish to be contacted by dealers and collectors. Please respect all owners and allow them privacy.

JOLLY GREEN GIANT

There's is a Panama Green Holden Overlander 4-wheel drive well known to the Victoria Police. The colour has been referred to as squashed bug green, puke green or just bright lime green. Its owner Joe calls it the 'Jolly Green Giant'.

Joe Henry can't go anywhere without his partner, Karen, finding out about it. She works for Victoria Police in Bendigo. More than once she's been told of the ute's sighting by Ops. It was once seen at a funeral in Morwell, but the police knew it came from Bendigo. It was once reported outside a brothel. Joe works in computer support and travels around a lot, and had parked innocently outside a house which, unknown to him, was a brothel.

The Overlander carries all sorts of computer hardware now, but it has had a varied life. It once worked in The Otway Ranges which is a cold, wet and windy area near the southern ocean in Victoria. Joe Henry used to be a farmer there and the ute worked in these very wet areas. It has been bogged up to the doors more that once. The ute also played a very different role during

the Ash Wednesday bushfires in 1983 when it was used as a fire unit.

Its rear tray has been well used moving Joe and Karen's furniture into their new home, taking rubbish to the tip, and carrying items such as a large boom spray, fencing materials, tonnes of hay and even a huge load of Karen's jams, chutneys and sauces! The ute often tows an 18-foot tandem trailer.

It has a 5-litre V8 motor with 400 turbo auto. Karen says they 'wouldn't part with it, as it's easy to drive and to park, and "old faithful" is very reliable.'

Joe first saw it in *Bush Driver* magazine and fell in love with it instantly. It was custom built by Tasmanian engineer Arthur Heywood, whose company, Vehicle Engineering & Modifications, made a range of 4-wheel drive Holden utes, vans and wagons for off-road enthusiasts. It had 395,414 kilometres on the clock when I saw it, but 'it's like an old axe with a new handle' — it's on its third motor and the auto has been rebuilt twice.

In 1977 a motoring magazine writer described this special edition Holden: 'the Holden Overlander 4-wheel drive is a health hazard. I've never driven anything which caused so many twisted necks, dislocated jaws and strained eyes. Wherever you go, people just stop and stare, even if they happen to be driving the other way!'

Karen's V8 Vauxhall Velox

Karen Schlein of Williamstown, South Australia has an unusual ute. She bought it in 1981, when she was a 17-year-old schoolgirl. She says she saw a vehicle at Tea Tree Gully, backed into a shed which was falling down all over it. She didn't know it was a ute when she knocked on the door to see if the owner would sell it. And she'd never heard of a Vauxhall Velox!

The owner was a man of 90, a retired farmer, and he said he'd have to ask $50 for it. Karen beat him down and got it for $40 — her friends gave her a hard time about that.

Originally the farmer used it as a milk delivery ute. Karen even found notes under the seat which said 'two pints of milk, please'.

'I wanted something different, which didn't surprise anyone, my parents included. I used to ride motorbikes and have always been individual. Everyone thought I was mad owning an old ute with no heater or demister. Not even window winders. You just pull the window down or up. My husband Jeff was always interested in cars. We've been together since I was 15 and he did most of the work restoring it.'

When Karen bought the ute it was 'derelict, full of hay, chickens and mice. The interior was wrecked, rear timber tray floor was rotted away, but it had no rust or dents. We got it on the road pretty quickly. Jeff's Dad had a 6-cylinder Vauxhall motor in the shed, so we put it in. I had a part-time job then so I needed a car quickly. We got some pressure packs of grey primer and sprayed it.'

'I've never had any adverse comments about being a female driving a ute, but had a few comments about what I've done to it, like "you've buggered it up", "it's a waste of a good ute", or "it's sacrilege". Mainly though, most people love it and think it's great, don't know what it is or say they haven't seen one for years.'

Karen showed me her 'ute diary'. She has kept a meticulous record of it's history. The first entry says 'spent $1.50 — new welsh plug'. She also showed me the original handbook, a sales brochure, and other pieces of memorabilia she has. The lounge-room had numerous enlarged photo's of her ute.

It is now onto its fourth motor, a 4.2-litre 253 V8 out of a VB Holden Commodore. Jeff's father is a mechanic and helped install it. Jeff is a fitter and turner and 'very fastidious'. She says it is a credit to him. They have been involved in a hot rod club, and Karen says her ute is 'cool' and 'great fun'. Whilst it isn't original, she wanted something different and 'did what I thought looked best'. She wanted a 'personal touch'. It has velour interior, Chrysler mirrors, disc brakes, mag wheels, and twin exhausts. It is painted bright 'Guard's Red', which is a Porsche colour.

'It goes real well and has heaps of grunt, particularly in low gears.

'It's been everywhere. I've even carried Father Christmas in it.

'I also moved all my gran's furniture from Kangaroo Island in it, onto the ferry and back to the mainland.'

Karen worked as a dental nurse for ten years. The vehicle had been her everyday ute for 17 years, until recently when she finally knew that with two kids she needed more room.

Again, Karen decided she needed something 'a little bit different'. She now drives a 1977 Valiant Charger. 'It was a great disappointment to me, to think that our lifestyle has changed and we needed a vehicle with extra room for baby seats. I cried the day I put the ute in the shed.

'But I'll never sell it. It's part of my personality' she says passionately. She is now a fulltime mother to Katelyn 4 and a half and Nathan 2. Star, the 13-year-old Kelpie–Dingo cross, from a station 300 miles the other side of Iron Knob is part of the family and 'is the best dog in the world'.

Karen says she looks forward to the day when she can give the ute to her kids. Young Katelyn has already picked up the language and says she wants to 'go cruising'.

John Williamson was born in the old bush town of Quambatook on the Calder Highway in Victoria. As he grew up his lungs would have been filled with the famous red mallee dust and the wonderfully distinctive smell of golden grain wheatfields on his family farm.

When I was a youngster, mine were filled with the unique smells of our century-old family grocer shop and the famous blue mallee eucalypt oil at Inglewood 200 kilometres further south.

It turned out that I had some things in common with this remarkable country singer. He, too, is a staunch and passionate Republican, loves the 'real' Australia and has spent his life dreaming of it and writing about all that is wonderful in it. I'd sing songs about the kind of things John does if I could, but my singing voice sounds like Neville Wran 'doing a Pavarotti' on a rough day — bloody awful!

I caught up with John at his place in Sydney. When I arrived he was down in the backyard in the shed moving vehicles around. John's got a big shed — stacks of boxes full of music gear. We spent some time there yakking about utes and swapping yarns.

Amongst the Toyotas in the big shed is one of John's favourites — an FX Holden ute, with number plate ULURU. Another of John's Holdens, an FJ sedan, is now on display in Canberra.

Much of John's business is 'on the road' which means touring Australia and performing concerts here, there and everywhere. That's where his Toyota LandCruiser ute, the two Toyota troop carriers and the six-tonne Hinu truck come in. Whenever John tours, a convoy of vehicles loaded with gear hits the road — so keep an eye out for MALLEE, DINKUM, MULGA, and EMU333. Give 'em a toot and a wave next time you see them on the road between Cairns and Sydney, the Alice and the Isa, or any number of camps in between.

After a photo-shoot we went indoors and he showed me his latest flag design. We discussed the Republic, utes, memories and a whole lot more over a brew, surrounded by

some of John's countless music industry awards — platinum records on the wall like you and I have Arnott's biscuits in a bickie barrel: Mo awards, ARIAs, you name it, he's got it. The 'extra' special ones he keeps elsewhere. John's mind goes back to other utes in other times — his fond memories of the little boy with the bare bum watching his father pump up the tyres on the family's ex-Army, 1940s Chev — a good farm workhorse.

His first car was a brand-new 64 EH Holden ute. It was his father's way of getting his eldest son to return to the farm. 'Come home and work on the farm and you'll get a new EH ute' was all the incentive John needed. It gave him a 'romantic feeling, a wonderful sense of freedom, my own radio at last'.

The smoky dark blue EH was John's 'romancing' ute, with 'foggy windscreen on Saturday nights', he says with a cheeky grin. He would drive 37 miles to Swan Hill to play Aussie Rules footy and see his girlfriend.

For the rest of the week, though, the ute returned to serious hard yakka, nearly always loaded down with farm supplies — half a dozen bags of seed wheat or bales of hay and often a couple of 44-gallon drums of fuel.

It was buggered at the end of its days, a real 'paddock basher', especially after a large wheat header ran into it.

In those days speed limits weren't much enforced between the towns, so it was 80 mph all the way. 'The tar roads buggered the bush in the end,' he laments. We agreed that a half-hour down a tar road often leads to a major country city, whereas in the 1950s you'd think twice about heading any further than your local town. With the coming of better roads and even better cars the writing was on the wall for small bush towns.

In 1965 John moved onto his own farm at Croppa Creek, between Moree and Goondiwindi, and Toyota LandCruiser tabletop utes became the farm workhorses. In 1969 John got an EH Holden sedan, the same year that he wrote 'Old Man Emu' and from then on life began to change as music took over his life.

Other vehicles came and went — he had 'all sorts of crap vehicles'. He even owned a Volvo (one of only 500 in the world, an 1800 ES Sportswagon).

In 1971 John got his first unusual number plate, EMU055. The same year a Falcon ute with a 302 V8 thundered the bush roads with John at the wheel, so powerful

'I needed a bag of wheat in the back to keep it down'.

John sold his farm in 1976 to concentrate on music and because he didn't want to spend the rest of his life 'driving a tractor in ever-diminishing circles'.

John took on the music business and has been extremely successful, but only after a lot of blood, sweat and tears, belief in what he was pursuing, and plain hard yakka.

With each new vehicle John associates a particular song or album. His 'Road through the Heart' album is linked in his mind with the white Ford F100. 'By 1985, I knew I needed a ute again and the F100 with 351 V8 was made into a 'troopie' ute, specially converted for life on the road.' The F100's number plate EMU333 was kept when the ute was traded in, and transferred to John's Toyota LandCruiser.

In about 1987 he was heading home from a trip when he saw an FX Holden ute on the side of the road near Wagga Wagga. The bloke had found it in a paddock in original condition and bought it. John had to have it. He parted with $3000.

'Bringing the ute home was hell! The motor was okay, but the wheels crabbed, it was a hell of a drive to get it to Sydney, believe me.'

John has since spent as much on it again, having had it fully restored. The motor

purrs, the interior is excellent, and the new paint job in its original 'Desert Sand' colour seems appropriate. He say he is not 'the total purist restorer'. He likes his comfort, so the wheels are a bit wider than the original.

The kangaroo mascot on the bonnet has an interesting heritage. John was playing at a folk festival in Port Fairy, Victoria, and found it in an antique shop.

The FX now sports the number plate

ULURU, a 'good balanced' name and a place of great inspiration for John. The song 'Raining on the Rock' from his 'Mallee Boy' album is about Uluru (Ayers Rock). Both the song and the album are still out there working for John.

'Number plates get moved around a bit as vehicles come and go,' he said. One Toyota troop carrier is DINKUM, 'a great Aussie name and name of one of his businesses — Dinkum Music Road Company. John bought

John 18 months

the Dinkum number plate just before the personalised number-plate craze in Australia.

The second Toyota troop carrier is called MALLEE, in recognition of his birthplace and still his biggest selling album 'Mallee Boy', which has now sold over 300,000.

The six-tonne Hinu truck that carries all the music gear is plated with MULGA, another great name for the Aussie bush where John spends so much time. John's wife Mary Kay also has her special number plate — WATTLE; he says she is the 'wildflower' in his life.

John originally owned a terrific number plate which was UTE, but the Road Traffic Authority 'sneakily got it back and re-sold it'. That's another story.

John's LandCruiser EMU333 has been specially adapted for life on the road, set up with raised tray storage underneath, a queen-sized bed, canvas canopy, table tailgate, and even a wood vice.

Radios in all vehicles help communication along the convoy. 'It's a bit of a rig.' Sometimes they stay at motels, but John prefers camping out in the bush, close to what he loves. He and Mary Kay head bush a lot, often to their retreat in the mountains.

John was preparing for a 13-concert tour of the east coast of Queensland as we spoke. His eyes sparkled when he told me of his extended trip to Hermannsburg Mission, near Alice Springs to spend time with famous Aboriginal painter Albert Namatjira's nephew and the

Aranda tribe. There are big plans to learn more of their life and music. He was pretty fired up about it and I could sense the importance he has placed on the trip. The Toyota ute will be John's home while he's away and he even described a special water system he was planning to install on it.

It was time to move on, John had things to do, and I was headed back to Canberra for more work chasing utes. We shook hands, promising we'd meet up at Winton, in outback Queensland at the opening of the Waltzing Matilda Museum.

I fired my ute up, John and a mate were moving vehicles again, as I headed off and out of Sydney, admiring the grevilleas and eucalypts as I went.

Mervyn Roy Greenwood knows what hard work is.

'I grew up as one of 13 children. I left school at 13 and went to work cutting wood in the forests. I used to ride a bike 15 miles out or more and then 15 miles home again at night. Some nights I'd be so tired I was barely able to ride. We used to hard cut box logs 7-foot long by about a foot thick. You'd get a good belting if you didn't cut enough.

'Actually, even as a kid of about eight I would help my father load wood onto a truck at night. All the

kids did, even the girls. Dad always had two jobs: one on the railways, and cutting wood until about 10 o'clock at night. Even when he was 76, Dad was still working a gold-mine, up and down shafts 30 feet deep. He was 90 when he died. Mum's 88 and still going.'

Merv grew up in Maryborough and spent many years in the forests cutting poles for the State Electricity Commission and the PMG (Post Master General), as well as cutting firewood. He married when he was 17 and his kids Carol, Sharon, Janie, Kelvin, and Gary are all grown up now. All but Kelvin married and Merv has 14 grand-

children. In later years he started work at a pump manufacturers, doing fitting work, and finished with the company 17 years later as a foreman. He has always liked to use his hands. Now he lives alone in semi-retirement, but mows a few lawns for regular customers.

That's where his trusty ute plays an important role. It is a pale green 1965 Datsun 1200 ute that Merv 'saw in a caryard for $1000 about five years ago. About two years ago I bought a Morris major motor at Kyneton for $50. It bolted straight in without any worries, and it fired up straight off and has been going ever since.'

To show me how good the motor was, he fired it up, and it went first go and certainly is not a chugger. The speedo shows over 100,000 miles, 'but that's probably the third or fourth time around the clock. It stopped working years ago'.

Merv says the ute now spends most of its time 'in the bush'. He is a keen prospector, and living in the centre of the Victorian goldfields means he has plenty to keep him occupied. Using a metal detector most days of the week, he explores gullies and mine sites and has found plenty of old coins and some small pieces of gold, which he sells. He says although they are only small bits, 'they help you survive'. He hopes to find 'the big one, one day'.

The ute goes where most others don't. 'It is so low geared it crawls into places you wouldn't believe. I get to where no other prospectors would bother to go. I've been caught a couple of times. Once I was in a huge gully and rain came, the track became very greasy and I couldn't get the ute out for two days. I walked home about four miles. Other than that it has never really let me down. I've had a couple of nasty falls in the bush, so now I carry a mobile phone with me, which is handy.'

MINI MAGIC

Henry Draper still holds the lap record for racing a Morris Mini Cooper 'S' at Phillip Island Racetrack. The 1991 record he set for cars in the 'up to 1300 cc historic class' stands at 1 minute 58.3 seconds — about 200 kph down the main straight.

Minis are Henry's business, pleasure and life. He started racing EH Holdens back in the 1970s at Calder Park Rally Cross and switched to Minis in 1975. He sold his truck business and bought Northern Mini Parts which he has operated for 20 years.

He's still racing, and says he has made friends in every capital city in Australia as a result of Minis, and he's raced at most tracks — such as Calder Park, Bathurst, the Australian Grand Prix in Adelaide. He has competed in the Targa Rally in Tasmania a number of times and this year his wife Roslyn will be his navigator.

He owns about 60 Minis including 35 in a paddock on his property. The rest are in his factory, along with racks and racks of motors, doors, seats, windows, and all known spare parts by the dozens, stored in boxes and on shelves.

Henry owns a rare Mini ute. It is 'the' ute for the business — a great PR machine. 'It's a real head turner.' The ute is used for some deliveries, but 'mainly to collect the lunches from up at the shop'.

This particular ute originally came to Australia as a bare body shell in 1978. It was bought for Henry's second-in-charge Barry whose wife, Sharon, drove it as her daily vehicle. Then a few years ago Henry bought it back to use as part of the business. He has even used it as the 'lead out' vehicle at Winton Truck Races, 'which looked funny being dwarfed with a big Mac truck driving behind it!'

Henry imports and exports. 'I've got a container going to England shortly with three Minis and a rare 1966 Mini Moke (the 80th ever made). He points to a Mini Moke, 'that one has to be dismantled and shipped to Hong Kong'.

In Henry's factory I met a young mechanic from England who was working on a Cooper. His father in England is also into Minis and he says that in England they call the ute a 'pick-up'!

'Nearly all the farmers in the district where I come from had Mini pickups, but most of them are rusted out because of the continual wet conditions.'

He is amazed at the 'ute culture' in Australia. He says he'd love to take a Commodore ute back to England. That would be something really different.

'Henry's Mini ute is the "flagship" for the business. Everyone twists their neck looking at it on the road. It has a 1300 cc motor in it — the normal ex-factory models only had 998 cc.

'It attracts "the big grin factor" when you use it.'

MORRIE MANIA

'If you can't beat 'em, join em.'

This was Kaye Parson's response to my asking what she thought of owning Morris Minors. 'I grew up on a farm and was interested in horses, knitting, reading and travel. When I married John ten years ago, it was a case of my interest in travelling matching his hobby of restoring.

'He needed something to do and when he suggested doing up a vehicle, I said to make it a Morris or a VW. He said he wanted to do a ute, so we started to look for a Morrie. I was more interested in the travel and social aspects of the Morris Car Club.'

Since then both Kaye and John have done a lot of travelling — and restoring. They now own a 53 Series II Morris Minor ute, a 51 MM Series II 2-door sedan, a 1960 Morris 1000 panel van, and their present project is a sedan that is being made into a convertible!

John and Kaye work at a large hospital, she on the switchboard and he as an orderly taking people into the operating theatre. They are both totally addicted to Morris Minors. Home is shared with the Morries and two dogs, Penny the shy border collie and Rastus the lovable Kelpie cross.

'We found the ute by pure accident. We went to the Bendigo Swap Meet looking for one, but didn't find one. We were driving to have lunch at a nearby pub, and there in a back street was the little ute with a FOR SALE sign on it. We were meant to have it! The guy was moving to Queensland and wanted $600, but we ended up getting it for $400 and it all started from there.'

John has been interested in cars and engines all his life. His father used to bring cars home, do them up and sell them. Kaye, on the other hand, never knew there was such a thing as a Morris ute.

'Kaye really got into it as well and was not afraid to get her hands dirty. I came home from work one day to find a pair of legs sticking out from underneath the sedan. Kaye was stripping old grease and rust off the chassis,' says John proudly.

Kaye was eager to learn. 'John would say, pass this or that, or I'd ask, what's this? I learnt from there. Now I can hold my own with the men. I'm worse than John when it comes to Morries now, I love them.'

John and Kaye are so attached to the little Morries that each has its name painted on it. The sedan is called Seemore ('you see more in a Morrie'), the panel van

Wobbles ('it used to jump all over the road until we aligned the wheels'), and the ute is called Smokey ('because of the Jaguar steel-grey paint job'). The convertible, no doubt, will cop a name when it is restored.

Seemore belongs to Kaye and she has done much of the restoration work on it. It has a motor which she says 'has got a mild cam and is slightly bored out. And the colour is Aztec Turquoise. I can keep up with the rest of the pack in it now.'

John says the reason he likes Morries is 'because they are so easy to work on, mechanically and body-wise.' They travel really well at a hundred kilometres on the highway and hold the road beautifully' he says. 'We tried the ute out once and it got to 140 kilometres on the Hume Freeway out from Albury.

Kaye says they each have their own char-

acters and personalities. 'You can keep them standard or add your own personality to them.'

Their hard work and careful budgeting for restoration has paid off. At the 1996 Nationals in Canberra they won 1st place for 'Modified B Series II Utility' and they also won 1st place for the 'Custom A 1000 panel van'.

The Morries are used to lots of travel. The ute has been across the Nullarbor Plain to Perth and back, and also to such places as Iron Knob, Alice Springs, Uluru (Ayers Rock) and Coober Pedy. It's off for a three-week tour of Tasmania. It is used to dust, flat country, hills, bush tracks, claypans and corrugations, as well as open highways.

One stressful afternoon 'halfway across Australia' at Kimba, they blew a gearbox — which had jammed in 4th gear. John had to keep the speed up. 'I couldn't stop at some road works, so ended up going back to the pub that night and apologising for boring straight through. The workers just laughed and thought it a great joke,' says John.

Kay cut in to say, 'Of all the

places to finally break down it was Kimba. There were two old Morries in town, so we bought a gearbox on the spot for $350!' A real steal in the middle of nowhere.'

John continued, 'We started at 1 p.m. and pulled the gearbox out of the ute, walked across the road and pulled the other we'd bought and then put it in the ute. We had no idea how it would go but it goes beautifully and is still in the ute. We finished at 6 p.m. that night as the rain began to fall.'

MUDBRICKER

I ran into Chris Spenser outside a post office. We met again later and relaxed in the shade of his garage with a cold drink during the heat of the day. Chris is a semi-retired school-teacher and librarian. He loves his utes, and 'grew up in the back of my father's Vauxhall Velox utility'. Years ago he sold his 400 cc Honda motorbike to buy his first ute. It was a Datsun 1200. He needed a ute so he could build his first home — a mudbricker. The Datsun proved a sturdy workhorse and many mudbricks were moved on its battered tray.

The ute had been a plumber's ute and had a hard life before Chris bought it.

'Its top speed was less than 80 kph, it used petrol like a camel drinking at a water-hole, and often stalled at traffic lights in hot weather. When it stalled I had to lift the bonnet, whisk the air filter off, throw a capful of petrol down the carbie, jump back in, restart, then off we went. Joe the Jalopy as my girl-friend called it had no heater, no radio and no way to lock the driver's side door except by wedging a block of wood to keep the lock pressed down. It had a dried goanna's foot in the ashtray for an ornament.'

Chris's ute was a colourful sight with its signs and stickers on the side, including the original owner's plumbing business signs. Chris's stickers voicing some of his philosophies were stuck on the outside mudguard. 'It was always popular with my students for some reason,' he says.

'It carried a dozen black sheep from Ouyen once. They were the basis of my breeding stock and I travelled all night to keep the motor cool and the sheep quiet.'

After building the mudbrick home, Chris bought a Mazda 1100, one of those tiny 'round the town courier wheelbarrows' as he described it. It was called Lewis. Chris was attached to the little Mazda, but even after installing a new motor it 'proved to be a lemon on wheels'. It did have a history though. In 1982 it was his wedding ute.

'Being white and in pretty good nick, it was chosen to be our wedding car. It suited our wedding on top of Green Hills outside Inglewood, with its backdrop of nearby Powlett Plains, and Mt Kooyoora with the sun setting behind it. It transported us both to the site and later to the reception a few miles away at Tarnagulla.'

In 1983 Chris's mother-in-law found him and wife Robyn a Datsun dual-cab with only 3000 kilometres on it. Robyn detested the colour — bright red. When they went to pick it up they were told another person had offered a higher price, but Chris managed to buy it and has had it ever since. He says he bought it on the same day that Bob Hawke was elected Prime Minister. Several years passed before the ute was given the name Desmond.

Chris says he always has it maintained professionally, so the original diesel motor is still in good nick. It has carried all sorts of gear over the years ranging from straw, dirt, firewood and, when building his current house, building materials.

It now has 405,000 kilometres on the clock and, as Chris says, 'It has offered the best of both worlds, rather than having a trailer'.

NATALIE

'I've got a girl who thinks she's a boy — she wants a ute!'

In a small country town everyone knows everyone else's business. It is very easy to become the target of gossips. Natalie Hillerman knows what it's like to be 'the talk of the town'. She dared to buy a ute.

I met Natalie at the Swan Hill Emu Eaters B & S Ball. She is a 24-year-old school-teacher.

When she went to buy her black VS Holden Commodore V6 'S' Pack, her father wasn't too impressed. He said to the car salesman, 'I've got a girl who thinks she's a boy, she wants a ute!'

It is a typical country attitude that girls don't drive utes. Even now she says 'most people think I must be driving my boyfriend's ute. He's got his own.'

Her boyfriend Travis Mitchell is a farmer from Woomelang, Victoria. He drives a white VR Commodore. They met six months ago at a B & S Ball at Pooncarie, New South Wales.

Her ute is special. 'I love it — it's great for carrying all sorts of stuff — my swag and gear when I go to B & S balls.' She has been to 15 balls so far. She spends a lot of weekends away.

Home is on the family farm near Balranald, New South Wales, but she works as an Art Teacher at the college in Wycheproof, Victoria. Often when Natalie heads home her ute is loaded with farm supplies, drums of chemicals and the like.

'Girls generally don't have utes. I like something different. No matter who you are, with a nice ute you stand out in the crowd.'

Natalie is quick to point out that to be female and own a ute, you have to be prepared to 'take the flak'. A lot of blokes gave her a hard time. 'Mum's been a good support, she encouraged me to do my own thing. My brother, Alex, is good too, although he did say I needed to slick it up a bit, so now I've got tinted windows and mag wheels. He's got a new VS Commodore ute. Dad's really good about it now too.'

I spoke with her father and he laughed when I asked if he was the man who made the comment about Natalie and her new ute. He admitted he wasn't too keen about a ute at first, but he says the change in Natalie since her young days has been amazing.

'As a child she would go with me into town if it was in the car, but not in the ute. She was very quiet and didn't go out much but her days at university changed her. When she'd come home she'd always wanted to take the family ute when she went into town.'

Some of her girlfriends have got utes now too. When she first turned up for school at Wycheproof, it was 'the talk of the whole town. All the kid's wanted a ride. They still do. It blew their minds; Miss Hillerman in her own ute!'

Owning her own ute has brought its rewards. 'Because I'm from a farm like them and I own a ute, they can relate to me and we get on really well. I have to hide my B & S ball stickers under the tarp, though!'

Natalie is fiercely proud of her ute and her free spirit status.

She bought her VS when it had done only 7000 kilometres. After 12 months it had 34,000 kilometres on the clock. The only thing, she says, is that she'd never buy a black ute again. It's too hard to keep clean on dusty roads. Her next ute will be white. And she is adamant that it will be another Holden. Natalie laughs and says, **'Beats a Ford any day!'**

OUTBACK COP

Streets are pretty well deserted except for one open garage and a café. Where does one find a ute story? As I drove along the street I spotted two police vehicles (one a ute) parked outside the large police station. I went inside and before long I was introduced to Senior Constable Owen Bray, a nine-year veteran of the force, four of those years stationed at Broken Hill. He led me to an interview room where I'd be doing the interviewing — a change for any policeman.

The Broken Hill offices of the New South Wales Police are headquarters for the Barrier District which covers about a fifth of the entire state. Most is outback land, much of it on dirt roads. It stretches from the Queensland border in the north, to 200 km south of Broken Hill, and to places like Wilcannia, Cobar, Ivanhoe, Tibooburra. Often investigations lead the police into remote areas. That is where the 4-wheel-drive vehicles perform their main role. Around town the police vehicles include the twin-cab ute that Owen Bray was driving with canvas canopy and lockable cage.

There are three general purpose duty vehicles, an anti-theft sedan, two detective vehicles (one of which is a 4-wheel drive), a physical evidence 4-wheel drive, and a stock squad 4-wheel drive.

The 4-wheel drive vehicles are important to operations. Road conditions up here vary from rough dirt roads with corrugations, to boggy roads impassable in winter. It is an offence under the local government act to travel these roads in winter. The conditions range from the sandy country towards Ivanhoe, to rocky country in the Sturt National Park.

The police advise people always to carry two spare wheels and more equipment than you think you'll ever need. You need to reduce speed at night greatly because of the large number of kangaroos on the road, while during daytime you have to watch out for emus and goats.

Police work here varies. Elderly people wandering off from elderly citizens' homes spark full air-and-land searches. Dehydration from the hot sun is a killer. There have been suicide attempts simply by 'going bush'. This is all on top of the usual range of urban and rural crime.

Policing this country takes good vehicles and good communications, and officers always travel in pairs. The vehicles they have at the station are Holden Rodeos, Toyota troop carriers, LandCruisers and Nissan Pathfinders.

There are five female officers based in Broken Hill in a total force of 45, and add another 15 more for other stations in the District. Two Aboriginal liaison officers help on work with the Aboriginal communities. When any two officers 'go bush', it could be for a number of days so they are always well equipped, with swags carried in the back. They have HF radios, and are in touch with Royal Flying Doctor bases, as well as the police stations. The stock squad often needs to travel to very remote areas; their record investigation to date involved the theft of 9000 sheep and 143 cattle.

It is interesting work. Owen Bray says: **'Once you get here, you really don't want to leave.'**

'PADDOCK BASHER'

Russell Hartley is a Goulburn Valley sharefarmer with his mother-in-law Grace Haines (see *Grace & the Honeymoon Ute*). He bought this small bottle-green 1964 Datsun 1200 about eight years ago for $100 from a man in Bridgewater, near Bendigo, Victoria. The man was a fuel agent and had used it to carry 44-gallon drums delivering all around the district.

It cost Russell about $200 by the time he did up the motor. It's no longer registered and spends its days on the farm. Russell uses a new Toyota LandCruiser on the roads. He built the Datson flat top tray himself, to make it easier to load and unload hay to feed the stock.

But it also has another important job. Russell's daughters, Jeannette 15 and Helen 13, drive it on wet days up the track to the farm's front gate to catch the school bus. They leave it there for the day, and then drive home at night.

Russell was more than happy to pose for a photo with the old Datsun. Whilst he no longer drives it every day, it still has a place in his world.

Like all farmers, Russell is a dyed-in-the-wool ute lover.

'PASS THE TEST'

Travelling over 32,000 kilometres to research this book, I checked out many pubs, as well as utes. One hotel stands out for friendliness and real outback Australian flavour. Some hotels claimed to be *the* pub for utes, but were a disappointment. One hotel had only two utes parked outside and one was mine. But the Silverton pub attracts people from all over the world. In fact, people from 126 countries have signed the book kept in the bar. There is also another very special book — a Ute Owners' Register.

The Silverton Hotel in outback New South Wales wins my award for a pub in the middle of nowhere with unique charm. A pub that has attracted people such Mel Gibson, Peter Fonda, Don Johnson, Catherine Oxenberg, Sam Elliot, Dawn Fraser, Dick Smith, Ernie Dingo, Normie Rowe, Bob Ansett, INXS, Steve and Mark Waugh, Mike Whitney, Bryan Brown, Chris Heywood, Syd Heylen, Maurie Fields, John Meillon and many other actors, sportsmen, adventurers, singers and artists.

All have competed to 'Pass the Test', a fun way to raise money for charity. Sydney footballer, Tony Lockett, failed the test. Each year, 12 charities receive the proceeds of the Test, a different charity nominated each month. Money has gone to such

worthy charities as Camp Quality, Royal Flying Doctor Service, Lifeline, St John's, Kidney Foundation and more. When I was there a man who had heard about the Silverton Hotel in his home country of Israel rode up just to try and 'Pass the Test'. We both succeeded. I can't tell you what the test is because that will spoil the surprise!

I know of a pub that claims to have sold 12,000 of its promotional stickers at $1.50 each. Your purchase of a Silverton Hotel sticker will help someone else. So far, the Silverton Hotel charity register shows total donations of over $25,000.

This is a real pub. This is real Australian country at its raw dusty best. Movie makers, the advertising industry, radio stations, newspapers and magazines love it. The walls are lined with photos of moviestars and movies that have been made in the area. These last include *Mad Max II*, *Razorback*, *A Town Like Alice*, *Outback Bound*, *Blue Lightening*, *In Pursuit of Honour*, *Golden Soak*, *As Time Goes By*. Many commercials have also been filmed at the pub, including Beer Makers XXXX, Eveready Batteries, Carling Beer (UK), and others. When you visit the Silverton Hotel, bring a photograph of your ute; it may end up on the wall next to Mel Gibson!

Want to see the real pub in the middle of nowhere land? Turn left at Broken Hill, follow the bitumen road out and within half an hour of cruising you'll see a small community of historic stone buildings from last century. Many are galleries showing the work of local artists and craftspeople. You can't miss the Silverton Hotel.

Tell them the Uteman sent you. Ines McLeod, the licensee and her husband, Colin, will make you welcome. Give them a photo of your ute, buy a sticker, help charity and see if you can 'Pass the Test'. And don't forget to sign the special Ute Owners' Register.

I saw a grey and maroon ute parked outside a garage in Molong, New South Wales. I wandered inside to see who owned it, and have a yarn. It never ceases to amaze me what you find in this country when you take the time to stop and look. Beyond the door of his garage is another world. A world of amazing car building and restoration. Top of the line stuff.

Three Rolls Royce vehicles were in the process of being rebuilt, and there were other vehicles as well, including Italia, Reo, Bentley, and vehicle names I'd never heard of. In Peter Lamb's office are photo's of vehicles he has restored or built for people.

To give you some idea of his craftman-ship: one vehicle cost $250,000 to buy and restore. A month after Peter finished work on it, it was taken to America and sold for $1 million.

Peter Lamb is a builder by trade, but 26 years ago he took up restoring vehicles. He spent five years learning panel beating and panel forming, and what started as a hobby became a business employing ten people. He decided to wind down a bit, so moved to the country, where he now employs one fulltime staff member.

He is kept extremely busy with many vehicle building or restoring jobs for clients from many parts of Australia. He also teaches students who come to his garage and, with his expert guidance, restore a vehicle as a project.

Peter has been a member of the Rolls Royce Owners Club for 32 years, and owned more than one Rolls, over the years. His 1962 XL Falcon is his work ute and he says it has been a great one, with a beefed up motor and four-on-the-floor. He's happy to drive it about though it's a far cry from the magnificent vehicles he has created over the years.

His business is Collectable Auto Restorations — and plenty of Vintage, Veteran, and Classic vehicles have passed through his workshop.

PLUMBER 'BAZ'

Barry Budd was 'on the job' when I saw him. He was working on some pipes in a trench. His nicely maintained 1993 Toyota HiLux had drawn my attention. Most tradesmen's vehicles are usually pretty scungy,

loaded to the hilt and the worse for wear. Not so the Barry Budd mobile. It's exceptionally clean and tidy. He believes in looking after his vehicles.

'It pays to look after them, good advertising, and good when it comes to resale. My ute is garaged every night. I bought it new. I had a Rodeo before this, but this is a 4-wheel drive which I need in the wet. I also use it for recreation. We go fishing.

'I have a canvas canopy which I put on, and my boat is carried on a rack on top. With the extra wide wheels and Overlander tyres, I can go anywhere. They are great tyres — they seem to keep the ute on the road and it doesn't bounce around much. It's also very easy to work out of. I tow a tandem trailer into the bush and cut all my own firewood. It's very handy.'

Baz does all sorts of plumbing and household work — gas, sewerage, roofing, spouting. He's been self-employed for 12 years. He is married to Desma, and has a daughter, Belinda 23, and son Daniel 18. Desma runs the general store in their small country town.

Just before I left Baz, I asked him whether he had a ute dog? And it's Possie, the little three-year-old Jack Russell terrier, who loves the ute, 'I can't get her out of it.'

The family ute can sometimes become a treasured item to be handed down through generations.

'Pop' was a farmer in Central Victoria who had an old grey EJ Holden ute. Usually at least two of his beloved sheepdogs — kelpies — travelled on the back of the ute and they'd bark non-stop as he drove along.

Pop could make a 'roll-yer-own' cigarette in just one hand between thumb and one finger as he drove along. His old gnarled and wrinkled hands had a strength that only years could bring.

Pop's grandson and my brother were mates and, as I was a few years younger, I'd usually get the short straw when it came to travelling in Pop's ute. His favourite trick with all 'young uns' was to grap your knee suddenly if you happened to be seated next to him. He had a vice-like grip!

Pop's cigarettes played a part in another of his favourite tricks. Without a word he would put the wick of a sixpenny bunger fire-cracker up to the cigarette still in

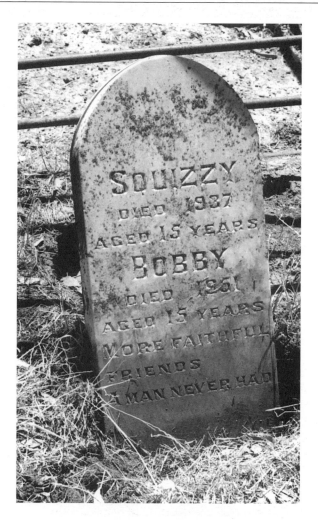

his mouth, light it and throw it at some poor dog running beside the ute. Many a yelping dog would run home from Pop's ute, tail between its legs, cured of chasing vehicles forever.

Pop really loved dogs though — headstones remain as testimony to that. Pop paid for professionally-made headstones to be erected over the graves of the sheepdogs he had trained on his farms. He gave me Skipper, my first dog, a great mate for many years. Every kid should have a dog.

Dogs loved Pop too, except those silly enough to come within reach of his cracker-throwing arm.

After Pop's death, the ute passed to one of his sons, AR, who used it constantly, and often loaded it up for the annual duck shoot at Lake Boort. Later it passed to Meddie, a mate, who collected firewood in it and took his gundog after bunnies on the blue mallee tracks out from town.

Then after some years, Meddie's son, Rob a tradesman painter, took it over. Meddie and both sons all have utes.

Pop's old ute spent its days loaded to the hilt with tin cans of paint and ladders. Then I saw it in the backyard.

Stuffed.

It had finally died.

'Queen of Speed' — Rachelle Splatt

THE WORLD'S FIRST 300MPH WOMAN

Rachelle is an amazing woman. Of her success as one of the top fuel dragster drivers, Rachelle says she simply has 'gone out there and done what I love. This is where I belong. With racing I've achieved a lot and as long as I enjoy doing it I'll stay. The day I don't, I'll quit.'

Spending time with the Rachelle Splatt Racing Team was interesting. It's both business and pleasure for them. Her father Ian and partner Wayne are part of her race team.

'When I first started racing in 1993 there was a lot of opposition. The men didn't want to race me in case there was an accident. Being female, it would look bad for the sport. A lot of the men said I couldn't do it, I wasn't good enough. Sometimes I would go home and I'd think that maybe they were right. Everyone was saying it.

'Then I'd get determined and say I was going to do it. The biggest thing was my father. Without Dad I don't think I could have done it. I'd go to him and he'd get me over it. He would simply say, "You can do it Rachelle, I know you can." Then I'd get fired up and do it. We're really close, Dad and me. Always have been. We are much more than just father and daughter.

'I still ring him every day in Queensland if he's not with us racing somewhere. When

I moved to the United States to race Dad even packed up and came to live there with me too. Now is the first time ever that we've been apart. He is running the Queensland part of his business Dragway Wheels and Harley Parts.

'There used to be a lot of egos in the racing business. Things are much better now. We are all mostly good friends. Once it was very "full on" but now we are pretty okay. I like to be friends with everyone.'

Rachelle first raced in top fuel drags in 1993. She became Australian National Champion that same year. She was the first woman to be licensed in Australia and as a result received a lot of TV coverage in the USA. She was sponsored by Luxor Casino to race in America.

On 3 March 1994 in Houston, Texas, she became the first woman in the world to race over 300 mph.

She lived in the USA for 12 months and raced there for six months before returning to Australia. She never had any problems in America as there were already other women racing there. 'I was just another racer,' she said. When Rachelle returned to Australia 'everyone accepted me. I'm equal now.'

She is much more than 'equal', as she casually puts it. 'She is the best thing that has ever happened to drag racing' is how one racer put it to me. 'Anyone in the sport will tell you she has been terrific, since the image of drag has never been as good. She is responsible for lifting the image. TV crews, magazines and other media all chase

whole day after racing.'

No driver has cracked the 300 mph on Australian soil yet. Rachelle has done 296 mph in Australia and thinks someone will get it before the turn of the century. She is having some time off but will return to 'really get back into it' later this year.

I asked where she saw herself in five years? There are some changes in the wind. As long as she is still enjoying what she is doing, she will be involved somehow in the racing business. She has achieved a status in her sport most can only dream of, so she deserves to go on to whatever she wishes for.

The ute in this story became very much a secondary issue: Rachelle, her father and the racing team were the inspiration. I was impressed with Rachelle even before I got to know her, and was struck by her friendly, unassuming nature. You wonder at the natural talent of this tiny woman as she handles such a powerful machine.

Her father summed it up when he said, 'She can come back from a near 300 mph drag and tell you exactly how the car

her for interviews. There was not much of that before Rachelle arrived on the scene.'

As we spoke, ABC-TV was setting up for an interview with Rachelle. Nearby, a group of young kids clutched colour posters and waited for her to autograph them. A group of men stood around, looking at Rachelle and her dragster.

Rachelle takes it all in her stride. She runs on pure adrenaline when at a racetrack, not only when she is driving. The demands on her time are exhausting. 'At the track it's always full on. When you are racing your body pulls 4–5G forces which drains you, and then you are always in demand with the public, crew and media.

'Sometimes people don't appreciate just how tiring it can be and you need a break in the trailer for a bit. I never really switch off until I get home. I usually sleep the

eight kilometres from where he lived. It had done 132,000 kilometres when he got it. And now? With a smile Ian says 'a lot!'

The 1989 Chev 454 with turbo 400 is now improved with an Allison transmission. They added the rear canopy as crew sometimes sleep in it. A second trailer like the one for the dragster is set up to cart quarter horses occasionally. So the Chev is a vital part of the Splatt family. A real workhorse.

Some of you purist ute owners may say it doesn't belong, that it is not part of Aussie ute tradition.

But what the hell!
What a rig!
What a woman!

performed. Then, when they have run their diagnostic tests, she would more often than not be spot on in her assessment.'

Rachelle has what it takes. She is the right stuff. She is an inspiration, not only to women but to all young people who want to achieve.

A race team like this costs about $280,000 a year to run. And that's if there are no breakages. For every pass down the track, a new clutch has to be installed. The rods last only ten passes, and a crank only twenty. One run down the track costs about

$3500, using about $600 worth of fuel.

Rachelle Splatt may now be a household name but she is very much all 'hands on'. She arranges all the airfares, accommodation, transportation, media, sponsors. I am reminded of what she said with that lovely cheeky smile, 'I've just gone out there and done what I love. This is where I belong.'

And what about the ute in this story? Rachelle's father, Ian, looked all over Australia for something big enough to tow the dragster trailer. Finally he managed to find the Chevrolet 454 crew-cab about

WORLD'S FIRST WOMAN
300 MPH
MARCH 4 - 1994

REO SPEEDWAGON

There's a magic ute living in New South Wales that I hope to see south of the border (in my driveway). Unfortunately, I heard of this rare vehicle just as the book was going to print. We've just managed to squeeze this page in because this is a real beaut ute. It is a tribute to owner John Rigby, who has shown great restoration dedication. He is a ute lover from way back.

The 1936 Reo Speedwagon is an extra special ute. 'Believed to be the only one existing and restored in Australia with the traditional Aussie ute style body manufactured by Mitchell Bodyworks in Melbourne. Americans call this an "express" body.' John knows of only six Reo's of this style existing in America.

Every nut and bolt has had attention in a meticulous process of rust-to-restored. John gave me a detailed eight-page summary of the work. As pieces were restored they were packed in cupboards; on top of wardrobes, even bumper bars under the bed! Finally after many years it is complete and has won several trophies for John.

The ute was originally purchased in 1936 by people from Colac. They bought it at the Melbourne Showgrounds where it was on display. It moved around a bit after that — John purchased it in 1991. He is the fifth owner. I'll be the sixth!

REPUBLICAN UTE

I was going to get up on my soapbox and tell you why we should become a Republic, but the photo of the EH Holden above shows that some people were interested in a Republic long before the current debate. My hero Henry Lawson wrote about it in the 1890s. The EH owner Geoff Hocking, was so angry when Prime Minister Gough Whitlam was kicked out of office that on 11 November 1975 he took to his ute tailgate with a paintbrush and angrily wrote 'Viva La Republica', and slapped a couple of union southern cross flags on it.

Geoff and wife Christine had been working hippies in England and Europe before returning to Gough's paradise a couple of month's before Gough got the flick.

Geoff's defiant stand brought instant responses. Often he would get a toot, a wave and a 'thumb's up'; some, however, were not impressed, including one bloke who wanted to rearrange Geoff's facial features.

Geoff's ute has a great history. Christine's uncle was a chook farmer at Eaglehawk, and his sons used 'the bush bomb' to run around the paddocks. Geoff and Christine were desperate for a cheap car on their return from overseas. For $500, they got the ute, which came pre-dented.

Geoff and Christine started to build their mudbrick cottage on the hill at weekends, and returned to Melbourne to work during the week. Christine's brother drove the ute around town all week, and would leave it on the block with keys in the ignition, ready for Geoff to collect at weekends, to carry mud, shovels, wheelbarrows, fenceposts and anything else they needed.

'It was a damned good workhorse. Invaluable,' says Geoff. Christine says, 'a ute's a ute'. For four or five years the EH worked in the construction of their new home, then the day came for it to go. Geoff spotted an ad for a Commodore Run-Out Sale less $2000 for any vehicle registered and going.

They bought a new 4-cylinder Commodore stationwagon for $11,000 and, when the car salesman saw the old bomb ute parked across the road as trade-in, he said to Geoff, 'And how much do you think that's worth?' Geoff replied, '$178 for the muffler I put on it, but for you today, it'll cost you $2000!' The salesman wasn't too impressed but Geoff produced the advert from the paper, and so the deal had to be done. They drove out with the brand-new Commodore.

A couple of months later they saw the old ute. It had been sold to a bloke who was down by the Yarra River, with a gas bottle in the back for blowing up balloons for kids! Where it died we do not know.

The angry young man and his EH is now nearing 'old codgerhood'. He and Christine have three kids, Alexandra 19, Nicolas 17 and Julian 10. I gave them a red heeler dog years ago, but it had the sense to move to a new home. Now, only chooks roam the hills and gullies on their patch.

Apart from the BMW, the Peugeot and the MG, there's an old yellow 75 Holden ute that the old fella drives, but it will soon be passed onto eldest son Nic as 'P'-plate time arrives.

ROADWARRIORS

They're everywhere
... the beaut utes
... helping to build our roads
to the future.

RR 'SILVER GHOST'

Some utes have great pedigrees. Even British Prime Minister, Sir Winston Churchill warmed the seat of this 1924 Rolls Royce 'Silver Ghost'. It started life as a 7-seater touring sedan, with a bright yellow and polished aluminum body by coach builder Park Ward, and installed with the rare Steven Grebel headlamps. It was specially built to order and lavishly appointed. It even had a gun box built on the running boards.

British politician Sir Reginald Purbrick was friends with Sir Winston Churchill. The car was shipped to Europe where Sir Reg and Sir Winston used the vehicle for 'motoring trips and holidays'. It stayed on the continent at Beauritz, the playground for the rich and famous.

In 1934, Eric, son of Sir Reginald, moved to Australia to run Chateau Tahbilk winery and vineyards. The car came too. He finally traded the car in to a local dealer in Shepparton in 1950 for a new Rolls Royce 'Silver Dawn', which is still in the family.

The dealer couldn't sell it, but finally in 1951 a local farmer, by the name of Clapperton bought it on the condition that he could trade in his Belgium-built Minerva ute, and that the Rolls be made into a ute, using the rear tray from the Minerva. This was done, and a new life for the Rolls began. It became a farm ute with two main uses. In summer it was the local district's fire truck, fitted with a water tank on the tray, whilst during the rest of the year it was used for normal farm duties.

During the 1960s it changed hands a couple of times, to people who had intended to restore it but never did. Then a Rolls Royce enthusiast, who was also a recluse bought it and it was put into a garage and left for many years, covered in dust with vines growing all over it. Rumours developed about the car and the recluse. Pilferers had broken in and various bits were pinched off the vehicle.

About 1988 it found a new owner in well-known Melbourne Rolls Royce collector and restorer, Robert McDermott, who got it in a shabby but fairly complete state. He put in a new battery, points and plugs, then started it up and drove it to a rally! Not bad for a vehicle that hadn't been on the road since 1958. Robert went on to fully restore the chassis and all mechanicals.

Robert McDermott's own background is the stuff books are made of. Son of a Melbourne businessman who ran a group of engineering and manufacturing businesses, Robert grew up around cars. His first car was an Austin 7, which he got when he was nine years old. He had his first Rolls Royce, a 1923 'Silver Ghost', when he was 13 years old.

He learnt a range of skills such as welding, design and drafting in his family business and became seriously interested in vehicles. He went on to do a degree in economics, then worked in the family engineering business, before starting his own Rolls Royce servicing and restoration business in 1975. It became the largest centre in Australia. He moved to new premises five years ago and remains a major a Rolls Royce repairer, with 30 staff, and he is a Suzuki dealer as well. Robert has the most amazing recall for detail and history when it comes to Rolls Royce.

Finally he decided to sell the Rolls Royce ute, and its current owner also needs a book written about him! John Wagstaff is a civil engineer, and he looks like finishing where Robert McDermott left off. A keen Rolls Royce collector, he purchased the ute because it was 'something a bit different and a bit of fun. Also the carrying capacity is a bonus.'

He says it cruises at 100 kph beautifully, but is heavy on slower speeds. It averages about 10 to 12 miles per gallon! He says it really is an excellent vehicle. When he picked it up in Melbourne, he drove it to Adelaide, then put it on the Ghan train to Alice Springs. From there he drove it to Uluru (Ayers Rock) in a rally, but on the way back it blew a hole in a piston, so it went home to Brisbane on a tilt-tray tow truck.

It is presently undergoing final restoration. A new rear tray has been built out of jarrah and other work was under way when I interviewed John. By the end of this year it will be completely restored — and I want a ride in it!

The ute is a long-term investment for John. He originally intended restoring it back to a car, but a Rolls Royce collector in England pleaded with him not to, as there are not many utes about. I think it could not be in better hands (except mine) as utes are very much a part of John Wagstaff's business. Of the 75 vehicles his company owns, about 50 of them are utes, mainly Toyota LandCruisers.

The company that he started with his wife, some 16 years ago, is now operating in most capital cities. They carry out civil engineering and foundation contracting, working on some of the biggest projects in Australia, such as the new Docklands Stadium and Federation Square in Melbourne, to name just two projects.

The 1924 Rolls Royce 'Silver Ghost' ute is a fitting advertisement for a man who has so many utes. When fully restored, both John, and Robert McDermott tell me it will be valued at about $250,000. Whilst he says he may have over-capitalised on the vehicle, I think it will be a ute long remembered, and what better publicity machine for promoting your business could you get? Not your average EH Holden ute with a sign painted on it!

RON'S CHEV

In 1981 Jack Evans saw an old 1949 Chev ute covered in rubbish in a fence-enclosed paddock. He thought he'd like to buy it to restore as a plumber's ute. When he managed to track down the owners of the land, they didn't know who owned the ute and said he could have it. Some time later Jack got sick, so a friend of his, Ron Davis, said he'd help restore it. Jack had cancer so handed the ute over to Ron, saying he'd watch him restore it, but he died three weeks later.

Ron has had it ever since. He restored it in 1988 but had trouble with the motor, which had been welded to the front end by a previous owner. And there were 'lots of funny noises coming from the rear end when I got it'. Now it has been fully re-engineered, a Chrysler Hemi 265 motor installed, and a Supra 5-speed gearbox. Ron consulted with an engineer but did the work himself and finished in 1991. He then built a special canopy for the back, and with his wife headed outback.

They joined an RACV tour to Woomera, Lake Eyre, Moomba Gas Fields, Innamincka, Tibooburra, Tilba, along the Darling to Broken Hill, then home via Balranald. In five weeks the Chev took them 7000 kilometres without a single problem.

After this taste of outback travel the ute headed bush again in 1993. This time it was eight weeks and 15,000 kilometres!

'It was a great trip which took us to Broken Hill, north to the Gulf, Lawn Hill National Park, Booroola, Katherine, Wyndham, the Gibb River Road, Windjana Gorge National Park, Derby, Broome, Rabbit Flat, across the Tanami Desert, Alice

Springs, Port Augusta, and home via Mildura. In all there were only two problems with the ute on the trip. The first was a broken main-leaf rear spring, and a rear wheel bearing, which we fixed when we found an HG Holden wreck on the side of the road — pulled the spring out and shoved it in the Chev. That is the best thing — having all Holden running gear on it makes it so easy to maintain. We had the wheel bearings flown out later to Katherine.'

Ron uses the Chev daily as a work ute. It is also used to tow a caravan, and since being restored has travelled 150,000 kilometres. Ron now does handyman jobs — gardening, mowing, painting, repairs and other work. His work background was as a farm manager for 15 years before joining the police force. He retired from the force after 25 years, rising to the rank of Detective Sergeant, officer-in-charge of a CIB branch.

Ron did courses in panel beating, spray painting, upholstery, welding and diesel mechanics. As a man who likes to restore motorbikes and cars, the courses have been a great help. He also owns a fully restored 1934 Ford coupé which he rebuilt over a ten-year period from the base up. The Chev ute will be in the garage for a while with

Ron's 57 BSA Golden Wing motorbike, plus a Triumph 500 and a Kawasaki 750 VN; because Ron and his wife, Ruth, plan to drive the 34 Ford around Australia towing a caravan next year.

Runt of the Litter

Mick Hermann has heard most of the jokes about the littlest ute of them all — the Suzuki 'Mighty Boy'. People have said things like 'stick it over your shoulder and burp it', and comments about it being a 'motorised esky' or an 'esky on wheels' — 'fill it with ice and you've got a mobile esky'.

But Mick is very enthusiastic about Suzukis. He's made a business of them for the last 12 years. He'd much rather drive his 85 Mighty Boy to work every day and leave the LandCruiser to his wife. The Mighty Boy gets about 12.5 kpl and 90 to 100 kph, more if you strangle it.

The first Suzuki arrived in Australia in the mid-1970s. (I know; I drove one of the first models in Alice Springs in 1976 when I got married. The local Avis agent gave it to me for the weekend to go bush as a wedding present. We went to a motel instead!)

Mick says farmers love Suzukis because in wet paddocks they are so light they don't get bogged. He carries out mechanical repairs, deals in new, secondhand and reconditioned parts, sends parts all over Australia, and even to Suva and the United States of America.

Mick has driven his Mighty Boy from Melbourne to Mildura and as far as Bairnsdale in East Gippsland. His other ute is a 1985 Suzuki Sierra.

In years gone by before there were utes, many people, farmers in particular, used to remove the rear tubs of sedans, and build a tray to convert the car into a ute. Now, decades later, many of these great old utes have gone through a reverse restoration process.

Colin Nesbit's son Rex bought a rusty old 1927 Dodge ute at an auction for $300 and planned to restore it. His father helped by towing the wreck home. Rex began to dismantle the ute, but who would have guessed what that would lead to.

Colin 'got the bug'. He ended up buying the ute off his son. This was Colin's first project; he went on to turn the ute back into a tourer. It's an unusual vehicle, having had only one previous owner who came from a farm near Minyip, Victoria. It is one of only nine ever brought into Australia, with an all-steel body and frame. Since its restoration it has done 19,000 kilometres.

Colin then went on to own and restore a 1924 Roadster, which he drove to Perth and back in 1979, and then in 1981 to Alice Springs, Mt Isa and home via New South Wales. He added a 29 'DA' Dodge sedan to the restoration collection, followed by a 38 Dodge sedan and finally a 64 Buick. Four other wrecks arrived, were stripped and sold. He also became vice-president of the Vintage Drivers' Club, and organised rallies all over Australia.

It is an example of what hard work, scrounging, learning new skills, and dedication can do. Colin has sold off all his vehicles now, except the original tourer. It is his pride and joy, and will always remain in the family. He says, 'Vintage restoration has meant that I've met hundreds and hundreds of people from all over Australia.'

Their careers are worlds apart but a common love of horses has been with 'Sarge' and Vanessa all their lives. Sarge got his nickname when he was a horse-breaker and stablehand at Flemington Racecourse years ago. His old boss, trainer Pat Courtney, would tell him to do a job, and he'd simply go and tell someone else to do it. They thought he was leader material, and for a joke his mother sewed some sergeant stripes on his shirt. The name has been with him ever since, and many people don't know his real name.

Now he does shiftwork as an aircraft loader for Qantas, but he and Vanessa live on 17 hectares (42 acres) on the plains near a mountain range. A new two-storey home is shared with their horses and two dogs. Six-year-old Harry is a border collie–golden retriever cross, the 'pound dog' they got the day he was to be put down. Their other mate is Spike, a four-year-old blue heeler who decided he had to pee on my camera equipment box as I set up for a photo-shoot. Thanks Spike! They are both great dogs, very friendly and Harry's main trick is to retrieve cigarette packets and deliver them to Vanessa.

Sarge and Vanessa are great friends with other ute owners, Rod and Karen (see *Versa-U-tility*). Rod painted their house for them.

Vanessa is a woman of the 1990s, hates 'female' jobs in the kitchen, and loves to be out and about. She laughs and says her workmates call her 'the bloke' because she holds a heavy-duty driving licence and also because she drives a ute. She is a dental therapist and drives a large dental van to schools all over Victoria. She grew up on a farm, and has been riding since she was four.

She bought a new V6 Mitsubishi Triton in 1993, which she loves. So much so, she won't allow Sarge to drive it! 'I love it, it's great for towing horse-floats, and carrying hay and stockfeed. It's my first ute. Three weeks after I bought it my horse reared and her hoof landed on the bonnet. The dent is still there. I won't ever have it taken out.'

Her horse is thoroughbred 'Equal to None' a former racehorse which she nicknamed Kay, a magnificent looking black horse with a beautiful quiet nature. Vanessa spotted her in a racing stable and it was five years before she finally was able to buy her.

Vanessa is as enthusiastic about her red Triton ute as she is about her horse.

'It really is fantastic, the first day I picked it up I had some girlfriends with me. Some guys at a set of traffic lights were admiring the nice new shiny ute. You should have seen the look of disappointment on their faces when they saw a bunch of girls in it. They thought it should have been a bloke's ute. I actually slept in it the first night I got it, I unrolled my swag in the back tray. It was at a polo-meet at Werribee Park. Sometimes I think there's more boy in me than girl.'

Vanessa is quick to add her ute is a 'show pony'. 'You can't put anything in it that'll scratch it.' Sarge has the workhorse ute.

Her ute mostly comes in handy for her horse work. She has a certificate in horse breeding, is a club level assessor, competes in Royal Melbourne Show, does dressage, adult riding club, and writes a monthly column on horses and horse gear.

'I go to a lot of winter horse shows. Everyone gets bogged. It makes you feel superior when you can put it into 4-wheel drive and just drive off!'

'**Some males still have the attitude, "just leave it there love, we'll move it for you". They get a shock when I do my thing, and they find out I have a licence to drive 20-tonne vehicles.** When I got the licence to drive dental vans for work, I asked to go up to the next licence level so I can drive horse trucks.'

When I asked Sarge if he also had a heavy licence, he replied, 'No, I'm happy to sit there and drink beer while the missus drives the truck.'

Sarge is a real man's man. He is no slouch when it comes to horses. He won the inaugural 'Man from Snowy River' horse race in 1996 at the Royal Melbourne Show. The large silver cup and his various ribbons are on display. ('I won the beer drinking competition after it too.')

Sarge drives a brown 1977 Ford F100, with 302 4-speed motor. He describes it dryly: 'The clock shows 205,000 kilometres, but it stopped working about four years ago and it's done about that many again since.

'It's about ready to die — it rattles and moans. It's about to go into the shed, then I'll rebuild it. Got to get a car to get to work, which will be a real pain, not looking forward to it. The ute will be done up again, new motor, new diff, suspension re-tensioned, new paint job.

'I got stopped in the ute the other day by a copper who says the bullbar is illegal because of too many sharp edges, so I'll have to alter that.

'It's been a great ute for legally towing a two-horse float. I've had around 30 bales of hay on it. She was sitting down a bit when I carried three tonne of rocks.

'We couldn't afford a tractor, so I borrowed a set of harrows and hooked them up to the ute, put an esky in the back, turned the music up, and drove around harrowing about 35 of our 42 hectares!'

Everyone laughs when Vanessa describes the funniest sight on the farm which is Sarge in the ute when it snows. '**Because all the rubber seals around the windows have had it, and there are other gaps, the snow gets sucked in and he gets covered. You should see him when he gets out, freezing cold, covered in snow and teeth chattering!**'

Sarge casually replies, 'But it's good when it hits the ice, it never moves or slides like all the other vehicles because of the big fat tyres. It just sits there. A ute is great. It's simply a sign of freedom to do anything you want.'

SHIRL'S 2-TONE MARK II

There's a maroon and blue 1959 Mark II Zephyr in the town of Wonthaggi that is extremely well known. Shirley Wade has driven it every day for the past 15 years until recently. 'There are a lot of people in the district who pushed it to get it started in the early days. It used to go through batteries,' she laughs.

Cars have been a big part of Shirl's life. She says she's been interested in cars 'forever'. Her mother always fixed her own car and that's what Shirl does too. 'My grandmother had a Consul. My grandmother and mother were thrilled when I married David. He always had the bigger and better cars.'

The Zephyr Mark II is one of only 500 utes made exclusively for Australia. Two of them have made their way back to England. The Zephyr has been off the road now for a few months. Shirl says that an incredible number of people ask her 'where's the ute?' It awaits restoration.

'When we got it in 1983 it was pretty tired though it had rust in only one panel. We pulled the rear canopy off straight away and in the first six months we gave it a new paint job.'

Husband David comments that 'it took ten minutes to get it up to 50 miles an hour back then! We bought three motors from Yea — none of them were any good. We found a Zephyr ute upside-down in the tip and stripped it down for parts on the spot, including the motor. We built a good motor for the ute, but it's pretty tired now.'

They bought it out of the *Age* newspaper on a Saturday morning from a policeman in Pakenham who had bought it from the original owner, Angus Robertson from Nar Nar Goon, the local busdriver who had bought it new.

They still have the owner's manual with his name on it. The Zephyr cost them $350.

Shirl and David sold their property last year and moved into their new double-storey home. The ute went with them of course. It has carried all sorts of gear over the years and has been loaded down with motorbikes, engines, V8 gearboxes, rubbish, furniture, groceries, motors, drum kits, amplifiers, and much more. 'It's carried the kids hidden under the tarp down to the beach. Put the ute into first gear and it'll go anywhere, sand never stops it.'

When Shirl got the ute it had done about 50,000 miles. The speedo stopped 'a while ago now' at 161,000 miles. The older folk of Wonthaggi enjoy it, for it brings back fond memories to see it in its original condition. The ute will never be sold, but it will be rebuilt. David says the motor will be first. 'Nothing's been done to the ute over the years so it now needs everything — motor, muffler, gearbox, repainting, new clutch. Everything!'

Shirl has a final comment: 'The Zephyr ute is special. It's part of me. Like a comfortable old shoe that's worn out, I've put it away for now, but I will bring it out again and restore it.'

Simply...
Peaches, Pears, Prunes
...& Wombat Hubcaps

When you travel between Cootamundra and Young in New South Wales there are many orchard signs along the road displaying all sorts of goods for sale. One sign in particular caught my eye as I was motoring along. At the tiny hamlet of Wombat, a U-turn led me to the home and garage of retired shire grader driver Norm Patterson.

When he retired five years ago, Norm was looking for a hobby. A friend suggested he collect hubcaps. Now Norm has over 1000 hubcaps from most makes and models. He gets them from wherever he can and people bring them in to him.

He repairs and cleans them and buys and sells all sorts. His display certainly is unusual. He's been the subject of newspaper articles and the odd visit from a film crew over the years.

Norm even has hubcaps that he can't identify ... yet.

In 1969 Lance Culpan and his parents started a metal recycling business, and some ten years later expanded into other premises to start steel fabrication and selling new steel. Today Lance runs the business with his wife Judy, who is the book-keeper, and 20 staff.

If it's made in steel, Culpans are into it: housing frames, factories and all sorts of jobs, big and small. The recycling operation is still working from the original premises, but it is the new steel business that keeps Lance and his staff busy — as well as the Stouts.

A 1971 Toyota Stout ute is still the work-horse. The ute was purchased brand-new and is now onto its second motor. Lance says he chose the Stout because it was cheap at the time, economical to run and reliable.

A friend gave him a second Stout for spare parts. This Stout, a 74 model, was previously used to deliver paint from a factory. It remained in a shed on his farm but the load capacity of 1½ tons made him decide to restore it as a second vehicle for the business.

Now he would be lost without two Stouts. They are great for steel deliveries, on-site welding, and many other jobs. Both utes are used to get to and from work by Lance and foreman, Keith Howden. Lance has also used one to go duck shooting with a couple of mates, carrying a boat on the rack and camping gear on the rear tray.

When I was working for the Department of Conservation, Forests and Land in the 1980s, we all avoided having to drive a Stout. Anyone who has driven one knows they are mongrels to drive long distances. But they are tough!

Squeezing three large men into a Stout is like putting two fat ladies and a great dane in a Suzuki Mighty Boy.

I love the look of it — the shape — the curves. Today utes are mainly straight and all plastic.'

Stewart Brown is a 20-year-old, third-year apprentice electrician, born in Essendon, and he loves his 1964 EH Holden ute. He works the suburbs and beyond for North East Electrical.

His workmates say that as an apprentice 'Stewie is always willing to learn and work.'

'Halfway through last year I knew I wanted a ute and looked around for a while. A mate had done up an EJ — I loved it. I started checking car magazines and found the EH. My workmates had seen it the day before in a street in the suburbs!'

His story is like that of many young men who see and fall in love with a 'pre-loved' vehicle.

'I had a few mates look at it. Some said I shouldn't buy an old vehicle but I didn't listen. I wanted it, bought it — and then the engine blew up! And then the steering box went. I spent a lot of time on the phone chasing parts all over the place.' He lights another fag.

'I changed the 3-speed manual box over to a 5-speed. Converting it was another problem, and where I got it done I ended up in arguments. Anyway I finally managed, and now it runs beautifully.'

A lot of old blokes in their 50s and 60s admire it, and say things like, 'Ah, there's nothing like that being built these days'. I always get lights flashed at me, or a wave. People really admire it. People who own them, also appreciate the way it is kept.'

His girlfriend of two years, Amy, 'is a real good sort'. She works as a customer-service operator for FlyBuys.

Like a lot young blokes Stewie is a bit of a lad. He works hard, and plays hard. His vehicle reflects his get-up-and-go approach to life. He likes music. Grunge music mainly.

'I really like groups like Regurgitator, Prodigy, Foo Fighters and Trance. I like music up loud in the ute.'

'And he likes to lay a bit of rubber as he leaves work,' one of his mates chips in.

Once Stewie was 'hooning' on a wet road and wiped out a keep-left sign on a round-about. He has learnt his lesson through his bank book — the hard way. What young buck hasn't?

The old red 186 motor got up to 120 mph. No more of that though. 'Maybe that's why it blew up!' he says with a grin. He's done about 13,000 kilometres on the new motor in the last six months or so.

Why a ute?

'It's handy to have, good for camping, to carry gear in it. I do a lot of skin-diving down the beach at Lorne. The family have certainly found it handy. I'm always moving something with it — fridges, wardrobes, drawers, couches. Move it here, move it there.

'And, yes, the ute's been christened!'

He locked the keys in the ute on New Year's Eve and it took the RACV man an hour and a half to get it opened. 'I felt sorry, he was just a young bloke and couldn't get it opened.'

Stewie is a real true-blue Aussie. In between work, his ute and Amy, he plays Aussie Rules Football for the Aberfeldie Gorillas.

Getting Stewie to sell the ute won't be easy. He aims to get it back to what it should be. 'If I can afford it, I'll get everything on it back to original condition — and hope never to sell it.'

SIGNMAN

If you are afraid of heights then you wouldn't want Vince Larocca's job. He installs those huge signs you see on the side of multi-storey buildings. His work has taken him up some of the highest buildings in Sydney, Melbourne and Adelaide.

Look at the top of Nauru House in Melbourne. Vince worked on the installation of the 'stars'. Other signs include Telstra signs 52 storeys up, and signs for companies such as Shell, BP, Target, Coles, K–Mart, Challenge and ANZ Banks, GMH and Goodyear.

As a self-employed sub-contractor to a Brisbane-based company, Vince needs to be mobile and self-sufficient. His 1977 Ford F100 is just the vehicle, he says. And he really loves it.

'Without a ute like this I wouldn't have a job. I supply all my own tools, ladders, conduits, wiring and equipment. I need plenty of room. I carry a lot of signs and scaffolding, harnesses and emergency gear. Sometimes, I tow a cherry-picker as well. By the time you put gas tanks on small utes like Holdens, you've lost half your space.'

Vince started working in an office learning drafting. He hoped to qualify as an engineer but hated being inside. He began an apprenticeship as a signwriter instead, but the money was no good. So when he had the chance to take on sign installing, he thought, 'Why not?' That was ten years ago. He now holds a Restricted 'S' Electrical licence, a Neon 'N' licence, a Travelling Tower licence and a licence for cranes. He has also done a lot of on-site training.

A family man, Vince is married to Tracy, and they have three girls, Jamie-Lee 6, Kara 4 and Jazzmine-Star, 18 months. And there's also Levis the maltese Shih Tzu. 'That's not a real man's dog,' I said. He laughed and agreed, saying he used to have two german shepherds, but they couldn't keep them where they now live.

What does Tracy, who drives a Magna, think of the F100 plastered with stickers? **'I don't know how he can drive it, it's so embarrassing. But it suits his**

personality — crazy. He's always been a bit weird,' she laughs.

Vince agrees, pointing out that there's a barber's chair in the lounge, and he also has an old petrol bowser and a phonebox. He likes to be different.

The F100 stands out. Stickers of every sort clutter the outside and all over the interior of the cabin. 'The kids love it, little Jazzie can hardly wait to get in and she never wants to get out. The stickers keep them amused on trips. People give me stickers wherever I go, and I also give a lot of stickers out to kids. All the local kids love it.'

The stickers started when one of his workmates made up a sign for the windscreen which read Larocca, and then another one was Larocca Signs, and it expanded from there.

'Let's face it. It is a work ute, and it doesn't matter what you do to it. I love my truck. Stickers brighten it up. I like to feel good. It's just like the guys who drive a Porsche. They like to look and feel good. It's no different.'

Vince has always had a liking for F100s. This one he bought three years ago from a mate at work. It was in reasonable condition. He added the mag wheels and roll bars.

'It's a man's ute.'

STUDEBAKER

There are two in his frontyard on the lawn and the backyard is full with 13 more (he once owned 26), plus as many parts as you want to count. Bob lives and breathes Studebakers.

'My wife, Joy, and I have had many arguments over the number of cars about the place. Still, she is a Studebaker enthusiast and was editor of the club magazine for many years. Each of our three girls, Debra, Michelle, and Christina have their own Studebakers as well, and one son-in-law is worse than me. Before my father died he said he wanted a Studebaker badge put on his headstone.

Bob was born in the Seychelles Islands in the Indian Ocean, north-east of Madagascar. His father owned a garage and was a self-employed mechanic. Bob did his apprenticeship with his father. His brother also became a mechanic. His father decided to move and his finger landed on Melbourne in an atlas. Everything was sold up within a month and the family of five kids arrived in Australia when Bob was only 18. He is now 63. He has worked as mechanic for the same business for the last 20 years.

The notion of a Studebaker Car Club was first mooted under the clothesline in the backyard of Bob Godley's place. He and his wife became founding members and have been in it ever since. The club celebrated its 30th anniversary in December 1997. They send out newsletters to about 500 members all over Australia, New Zealand and even the USA.

Bob's trade came in handy when he took an angle grinder and cut a 1964 Studebaker Cruiser sedan in half. 'It was ready for the tip. It was all rusted out in the back half. Everything needed doing on it: motor, gearbox, front end, the lot,' he says. 'It was given to me by a bloke who knew I collected Studebakers. I didn't really want a ute, but when I cut it in half, I thought instead of throwing it away, why not make something with it.'

Bob did all the work himself over an 18-month period, and the end result was a unique vehicle. It has changed a bit since then. He's replaced the timber tray with aluminium, and it now has a different canopy. When it was first restored in 1979 his hard work paid off. He has received many awards for the ute, such as 'Most Popular Vehicle', 'Best in Modified Section', 'Best Custom', and more.

It runs on a Studebaker 259 motor, 195 bhp, with a 4-barrel carbie, twin exhausts, overdrive, geared very high for cruising, and the heater is called 'an acclimatiser'. Joy did all the interior work, breaking many sewing-machine needles along the way. The ute has carried them to many shows and rallies in Adelaide, Broken Hill, Melbourne and Sydney.

'You name it,' says Bob. 'It's never let me down. They're so simple to work on and, when you do them up, they go on forever. Everyone thinks it is factory-made but it's purely my own design and I did it all myself. Everyone wants to buy it but it won't be sold. No, never. It'll always be in the family. It's a real eyecatcher.'

I wonder what Studebaker Bros would think if they came back now and saw what Bob has in the backyard. Henry and Clem Studebaker pooled $68 in 1852 to start a blacksmith shop in South Bend, Indiana. But it was J. M. Studebaker who transformed Studebaker Bros. He was working on the goldfields, but found he could earn more money by making and selling wheelbarrows to the miners. He joined the company and they expanded greatly, building wagons and carriages. By 1877 there were five brothers in the business. The company rose to great heights but ceased manufacturing in 1966.

I had to ask about all the Studebakers in the yard — what were they? 'Well now, let's see. There's two 1946s (including a rare Champion), four 47s ('including one rare one'), one rare 48 landcruiser, a 55, three 63s (including a stationwagon), the 64 ute, and three 65s (including a stationwagon).'

And Bob loves them all.

STYLEMASTER

Parked right up the back of a shopping centre next to my ute when I returned from shopping was a bright — I mean really bright — purple ute. Here was a ute story coming to me. I waited for the owner to return. Adam Holland 27, is a shift worker, working as an aircraft loader for Australian Air Express. His ute is a 1946 Chevrolet Stylemaster.

'It's different, I can't go anywhere without people asking questions. People even pull up beside me on the road and photograph it while I'm driving.'

Adam was looking for something different through the various car magazines and saw it advertised. He'd seen it before but was not financial enough at the time. This time he rang the owner, who posted a video of it to him. He then flew to Tamworth, took it for a test drive, parted with the money and drove it home!

The list of work that this ute has had done to it fills pages. It has a 350 Chev motor and 350 turbo.

'People have said I could have got a new car for the same money, but I wanted something different. I added a stereo, and an alarm system to it. It won the 'People's Choice' award at one show, and was in the Top 5 at another.'

Adam now goes to a few hot rod shows around the country. He's been to Narrandera, Canberra Nationals, all over Victoria, New South Wales and parts of Queensland. He's not looking to sell it, but to get further into the hot rod scene.

'Nothing gets thrown in the back. I went camping in it, but put foam in to protect it all. Something likes this makes you very much more aware of the road conditions, being so low you have to be careful of driveways and road humps. Because of our

sloping drive, it takes me five minutes and special planks to get it into the garage. Dad loves it.'

Adam's daily vehicle to get to work is a HR Holden, although he does take the Chev at least once a week.

Adam's girlfriend, Lisa, doesn't like all the attention the Chev attracts. 'She prefers to remain unnoticed. People are very careful about the ute. They never lean on it but put their hands behind their backs and just look it over.'

We got down to business and took some photos. Dezel the friendly rock-collecting pooch also got into the act with one of his rocks in his mouth.

'I went and tried it out at Calder Park Raceway once. For a pretty heavy [1600 kg] ute, travelling the quarter mile at 13.4 seconds — about 104 mph — wasn't too bad, I guess.'

SWAMP DODGE

Finally Bill said 'No, I'll give you $500 and I'll keep the bloody battery as well.'

So the deal was done. Three weeks later Bill's mate dropped dead. 'But, he'd already cashed the cheque, and the ute has been here ever since. Bill has had his money's worth. He drives the old white 1974 VJ Dodge everywhere. It is his daily work ute. He keeps it in good nick and hasn't spent much on it over the time he's had it.

Bill and his wife run The Bakers Swamp Art Gallery from their 101-year-old stone homestead, on the highway at Bakers Swamp, New South Wales. Bill is recognised as one of Australia's best-selling painters. When you walk into his gallery, if you are an art lover like me, you immediately want to reach for your credit card.

Fortunately I don't carry one, so my dealings extend to what cash is available. Sadly I did not add to my collection on this day, but I do intend hanging a Bill O'Shea original on my walls soon. His work typifies all that I love about Australia. Many of his paintings include woolsheds, mining scenes, shacks, timber mills, old towns, pubs and much more, including old utes on dusty and red-earth landscapes.

Call in and say hello and tell him I sent you. Don't forget your credit card or cash. Have a gander at the $500 Dodge while you're there.

A mate of artist Bill O'Shea said he was giving up installing and fixing windmills, and that he no longer wanted his ute. Bill asked him how much he wanted for it, to which the reply was 'you can have it for nothing, but I want to keep the battery'.

Bill knew it was worth money for him to have a ute, so his wife could retain the BMW, so he said he'd buy the ute. 'Make it a $100 then, but I keep the battery still' was the reply.

THE BLOODY MONGREL

Anyone who went to Ivanhoe Boys Grammar School during the 1960s knew Warwick Gregory as 'Wazza'. He was a couple of grades ahead of me, a mate of my brother, John. They have kept in touch ever since, although it took a few years for 'Waz' and I to cross paths again.

Wazza has had a few nicknames — 'Wocka', 'Ducks' ('my cricket wasn't too good!'), 'Sticks' ('skinny legs when young'), and 'Jackson' ('by an old boss'). But now, if you go anywhere near the town of Murchison and ask for Mongrel, there's a fair chance you'll be directed to his garage.

'Whenever I go in a chook raffle, or raffle for a slab of beer at the pub and in the footy tipping each year, the only name that's put on my ticket is Mongrel. They don't need anything else to find me.'

Five years or so ago he bought a speed boat and named it 'Bloody Mongrel'. And his white XF Falcon ute has a sticker on the rear window which says 'the Bloody Mongrel's ute'. Naturally.

Wazza, I mean Mongrel, began his motor-mechanic apprenticeship in Nagambie, where he ended up staying for nine and a half years before going into a partnership in a garage in Murchison, which they called Gregory & Monahan Motors. With his partner Owen (Mickey) Monahan, the business has been operating for 23 years.

The Gregory family are Ford people. Like his father, **Mongrel has had only one Holden and swore he'd never have another. It was an FJ which was 'more bloody nuisance than it was worth!'** He did court Gail in it though, and she became his wife and mother of three daughters, Narelle 23, Karen 20 and Shannon 18.

As I interviewed Mongrel and his father they rattle off Ford models that they've had: XK, XL, XP, XT, XB, XC, XF, XA, XM, XD.

And more. All their vehicles have been bought at Nagambie Motor Garage, where Mongrel did his apprenticeship.

Mongrel's wife Gail drives a 1968 Mark 2 Cortina, and there's another Falcon, a 1975 GS sedan in the front carport. His treasured 1968 XT GT Falcon Sedan (white with black vinyl roof) is in the rear garage and is awaiting restoration before his 50th birthday. We spent some time in the backyard taking photos of the ute with Mongrel and his faithful companion, the seven-year old bitch border collie–kelpie cross Murphy. Wherever Mongrel goes, so does the dog,

including to work at the garage every day.

The white XF Falcon ute looks great, very neat, well looked after, and a bit distinctive. It is a ute made up of Fairmont Ghia options that Mongrel fitted.

'The only thing that's missing is a limited slip diff, and I've got one down the garage ready to install.'

Where did it all begin for Mongrel? He grew up with a love of utes. The family never had cars until much later on, always utes. He learnt to drive in 1960, when he was 12, in his father's brown XK ute, one of the very first utes in the area — everyone else drove sedans.

His father told me that he never did any damage to his ute — only his wife and son. Mongrel confessed to rolling his father's ute.

'I remember coming home and Dad was in the bedroom getting dressed to go out. I walked in and said, "I've rolled your ute!" He said, "You've what?"'

Mongrel and his father both laughed, but I wasn't told what else was said at the time.

Life without a ute would be difficult for the Gregory family. Mongrel cuts and carts wood for two households, his own and his parents. Caravans, trailers and boats also have been towed. (His boat has a Holden 308 in it!)

Years ago Mongrel and his mates put money in to buy a pallet of beer from near-by Murchison East pub.

'It took the ute two trips to bring it all home! That's a lot of beer, but it's a good way to buy beer price-wise.'

THE FLYING SCULPTRESS

Bernice (or Bernie) was an air hostess when they were still called that with Ansett Airlines. I was an air traffic officer. We went out a few times but she was already in love with a man who was overseas. He later became her husband and the father of their two boys. I moved to Alice Springs and eventually married Janette, another of Reg Ansett's air hostesses. Reg Ansett and I were born in the same town, but he made more money than I!

I ran into Bernie in the local milk-bar after some 20 years. She lives up the track. We've had some great yarns since then, remembering the early 1970s.

She went back flying after her two boys grew up. Jon 20, is studying at uni and working part-time with Qantas. Paul is 17 and still studying too. Flying is in the family. Bernie is a flight attendant with Qantas now. Her husband, John, recently left Qantas after nearly 30 years. He has his pilot's licence now, and an ultralight licence, as does one of their sons.

She and John bought a Falcon Longreach ute at a government auction and it's in great nick. Jessie, the young pup, loves it. They live on a couple of hectares and run sheep so the ute is always in demand.

'Being in the country, a ute is so handy. We've carted soil, hay, black-faced Suffolk sheep, render for the walls of our mudbrick house, chipbark, ceramic pots, chairs to BBQs, and much more. John added a new tonneau cover, and a bonnet protector. We drew the line at mats — after all it's a bloody ute, it is meant to be used.'

Bernie has a great passion. She is a sculptress. Bernie's sculpting gear also fills the back tray: blocks of plaster, tools and the like. She travels to private studios for tutorials, as well as currently studying for her Associate Diploma of Visual Arts. She longs for the day when she can spend more time testing her artistic and creative talent instead of winging her way across Australia.

'I wouldn't be without a ute now,' says Bernie. John says the ute is great in the bush, but they wouldn't need one in the city. A spirited debate between Bernie and John ensued as to whether they'd need one in the city or not, and there is no doubt that Bernie is a ute girl.

'I always liked the idea of a ute,' she comments. 'My lifelong dream was to live in the country, to own a ute and a dog. Now I've got it all. The loveliest people I've ever met are country people. Childhood memories are the best — utes, lovely people, salt-of-the-earth people.

Bernie is the granddaughter of Charlie Gill (see *Concrete Charlie*).

QUEENSLAND...
AUSSIE UTE Spectacular

McCullough Signs

THE GUMS SCHOOL QLD.

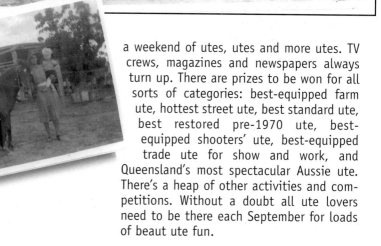

The tiny community of The Gums in out-back Queensland, 120 kilometres west of Dalby, hosts the Queensland Aussie Ute Spectacular and State of Origin Challenge in September each year. With a population of next to none, there is a total of only 18 kids at the little school. The ute competition raises desperately needed funds to keep the tiny school going. This year organisers expect a huge turnout with utes coming from many parts of Australia.

Ute competitions are not new; organiser Karen Stedman admits her committee pinched the idea from New South Wales. But they certainly like to make their event a weekend of utes, utes and more utes. TV crews, magazines and newspapers always turn up. There are prizes to be won for all sorts of categories: best-equipped farm ute, hottest street ute, best standard ute, best restored pre-1970 ute, best-equipped shooters' ute, best-equipped trade ute for show and work, and Queensland's most spectacular Aussie ute. There's a heap of other activities and competitions. Without a doubt all ute lovers need to be there each September for loads of beaut ute fun.

T-MODEL JACKAROOS

Ben Chaffey was a man of vision. His father and uncle were the inspiration for the founding of the irrigation scheme which changed life on the Murray River at Mildura and Renmark.

As Sidney Kidman was king of the cattle industry, so Ben Chaffey could be called the king of the sheep industry. Chaffey owned many large sheep stations in three states and saw himself as the largest sheep baron in Australia. Everything he did was on a grand scale. He had many top racehorses as well.

As part of some historical research I was doing into the Chaffey dynasty I interviewed many people in Australia, Canada and America. One was Alan McCalman who had worked on one of Ben Chaffey's properties during the 1920s.

In 1926 the property where 16-year-old Alan McCalman worked, one of twenty owned by Chaffey at the time, was a 330,000-hectare (823,000 acre) station. It employed nearly 260 men and women and even had its own railway line. Every year they shore 101,000 sheep. On all his sheep stations Chaffey installed power systems, refrigeration, lighting, septic systems. He had his own phone lines to the property and then out to all the various outstations so jackaroos could be in touch with the homestead.

Irrigation was another of his innovative schemes for dry areas inland; he intended to own properties from Queensland to Victoria and make his empire drought-proof. And all his stockmen were taken off horseback and put into T-model Fords which were converted to utes.

Ben Chaffey was a big buyer of Fords: numbers purchased varied from 18 to 23 annually. Documents show that Chaffey's pastoral company paid £315 16s for each ute (including insurance and freight from Ford Geelong to Dove & Chaffey, Ford dealers, Real Estate, Stock & General Agents in Mildura).

On the property near the Darling River in New South Wales where Alan worked, Chaffey would send three of the utes down to Mildura every six months, and three new ones would be sent up and handed out to jackaroos or whoever was due for one. These T-model jackaroos were a new breed.

On this one property, even 16-year-old Alan McCalman, who was the assistant bookkeeper, had a ⅓-ton T-model ute to do his extra duties checking the homestead windmills. Chaffey also bought 1-ton Fords for heavier loads, and two Dodge utes for the station manager and Chaffey Company boss. Chaffey himself travelled from property to property in a chauffeur driven Rolls Royce.

Most graziers had petrol delivered in drums, but Chaffey installed underground tanks and had his own petrol pumps installed. He was always ahead of his time.

One of Chaffey's properties is now in the hands of Jock (see *Jock's Mercedes*).

The droughts of 1927–29,

followed by the Great Depression, ended up ruining Ben Chaffey. Some of those I interviewed left no doubt that Chaffey would have succeeded again but he died before he could rebuild his empire. He managed to keep his Rolls Royce though right up until the day he died.

If Chaffey were around today you can bet his stockmen would have gone from T-models to utes to motorbikes to light aircraft. This visionary Australian would have led the way, and loved every minute.

TONY & DEIDRE'S '54 FORD F100

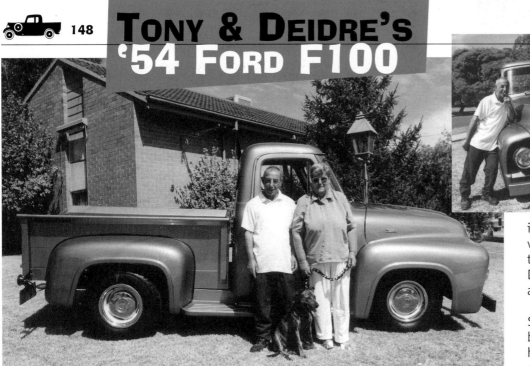

an FJ Holden sedan in the backyard as well. His son is also into hot rods. The F100 is entered into Rod Shows. It won 1st prize in the Centre State Rodders Pocket Run; it won 2nd prize the week before I interviewed them and then another sash and trophy just after our interview. Tony and Deidre and Tess, the bull terrier, travel a lot and love the Show 'n Shine Shows.

Over the past seven years a total of $10,000 has been spent on the ute 'and a bloody lot of hours'. Tony says every time he pulls in to buy petrol, what should be a quick job turns into a lengthy one, with people looking and asking questions. He could have sold it many times.

Tony is a bus driver, now on town-route driving, retired from the long-haul 'all over Australia' bus touring. He's been 'on the buses' for ten years, and before that was a train guard for 34 years until being made redundant.

The 54 Ford F100 is his first ute — not the usual red or yellow F100 ute but a snazzy metallic Polynesian Green one. And what a eyecatcher it is.

'**I just love the shape of it,**' was Tony Pianta's reply when I asked why he restored his 1954 ute. 'I always wanted an F100. **I thought I'd restore it in six months — it took a bit over seven years!**' he laughs.

'It was advertised in the local paper in 1990. I drove to Maryborough to see it. It was in very poor condition, a rust bucket. It had been a farm vehicle. I paid $1100 for it.' After much hard work Tony finally got it on the road about six months ago.

'We rebuilt it from chassis up, new skins on the doors, reupholstered, rebuilt a motor for it — it now has a 302 C10 auto. The steering and axles are original. My son Rick is a welder and he made a new rear tray for it. It's also got disc brakes on the front.

'The spraypaint job was done by Shane Rowe, of Southern Rod & Customs in Deniliquin, New South Wales. He only does hot rods.'

Tony drives the ute to work each day. He has an 'S' series Valiant to be restored and

I spotted Wayne Gunn's ute on the side of the road. He was riding a lawn-mower around a Lions Club park in a small bush town. Wayne is a nice quiet bloke and he is happy to talk about his secondhand ute 'Top Gun'. He is proud of how it has turned out.

Because Wayne has been interested in 'truckin' since he was 18 years old 'Top Gun' reflects the trucking influence. He has even installed large air horns from the latest Mercedes semi-trailer. His own work involves carting livestock, garden-ing and lawn mowing. His ute has purpose-built ramps so that he can ride his lawnmower straight onto the tray.

The stock crate allows him to carry pigs. In the true 'multi-purpose' style of ute ownership, Wayne also uses it to go camp-ing. It is fitted out with a sunroof, stereo system, UHF and CB radios.

VANGUARD

**Sunday afternoon.
SHEEPHILLS.
Outback Victoria.**

I saw the sheep but stuffed if I could find the hills! It is sheep and wheat country in the Wimmera — not many decent hills out here. I found a ute though. The story goes that a German farmer originally owned this 1949–54 Vanguard 'Standard' ute. It was British-made but this one was assembled in Australia; it still has the kangaroo bonnet emblem attached.

The farmer sold it to a bloke who was a rabbiter from Sheephills, between Minyip and Warracknabeal. The rabbiter used it for spotlight shooting, with a 'special built' rack on the rear tray to shoot from. The man was a bit of collector of all sorts of things. Finally, in 1993, the day came for him to clear out the shed and have a clearing sale. The ute was one of many vehicles — old Holdens, Chryslers, Pontiacs — to go under the hammer.

When the auctioneer came to sell the Vanguard, Geoff Niblett was in the crowd. He liked Vanguards, previously owning a 1960 Vanguard 'Estate', a 1959 sedan, and then a 1962 stationwagon. He says the ute was the best vehicle on offer that day and he wanted it.

'Bidding started at $20 and before long half a dozen or more blokes were bidding for it. I really wanted it and finally it was knocked down to me for $100. One bloke in the crowd clapped and yelled out "Good on ya' Nibby", says Geoff with a laugh. He had plans to restore it.

Geoff's wife, Debbie, continued the story. 'He didn't bring it home that day and wouldn't tell me what he'd bought for awhile. When he finally did tell me, I said "couldn't you have got the junk out of the shed first? What are you going to do with it — leave it out there to rot?"'

Well Geoff didn't restore the Vanguard, and it's now been sold to join the stable of utes in the *Beaut UTES* Museum. Nowadays, this model may be seen as a bit of a 'dag mobile', but in their day they were pretty reliable vehicles.

Geoff liked them because they 'were a good old unit — easy to work on enginewise. A good workhorse.'

There is a good reason why he never got to work on the ute. He's a real Holden man. Locked away in the shed is a magnificently restored and award-winning FJ Holden sedan. He is now restoring a FX Holden ute!

Debbie is also very interested in Holdens, but she'd really like a Chevrolet. 'I drove his Holden once, and don't want to ever again. Not interested in Vanguards either; would rather have kids any day!'

Geoff and Debbie are parents to Kylie 11, Mandy 9, Jessica 5, Brendan 3, Daniel 2 and Teagan 6 months. The other family member is Kizzie, the three-month-old Jack Russell–fox terrier pup.

Geoff is a former bulldozer driver of many years, but is now a maintenance man at the local primary school. He is also a keen restorer of stationary engines and is a member of the Dunmunkle Sump Oilers (Machinery Preservation Club) and the Holden of Age Club. He is also an active SES (State Emergency Service) member. His brother, Terry, is another a Holden man; he owns what I love and I need in my life — an FJ Holden ute!

VERSA-U-TILITY

VERSATILE: *Variable or changeable. Many sided in abilities. Capable of or adapted for turning with ease from one to another of various tasks.*

One of the joys of ute ownership is that the vehicle is versatile. Karen and Rod are a painting partnership specialising in decorative applications and painted finishes. They travel a lot in their job to places as far apart as Brisbane, Sydney and Melbourne, including paint contracts for the Harvey Norman chain. With Karen's background in fine arts, television and stage painting and Rod's tradesmanship they are kept busy.

Karen previously owned a 'champagne' (bronze-gold) coloured XC ute that had been 'hotted up' but reluctantly gave it up for a sedan. Rod owned a Valiant ute with a V8 which, after many months of restoration, was stolen from their home and wrecked.

Now they are both Falcon owners, a sedan for leisure and a trusty cream XC ute which is a seven-day-a-week workhorse. The ute has a throbbing V8, runs on gas and has numerous extras. With 450,000 kilometres on the clock, it now has a new motor, clutch and gearbox. It will be complete when some bodywork is carried out.

Loaded to the hilt is its usual state: large drums of paint, ladders, dropsheets and half a tonne of bits 'n pieces, and that's just the 'tradie trailer', let alone the tray of the ute itself!

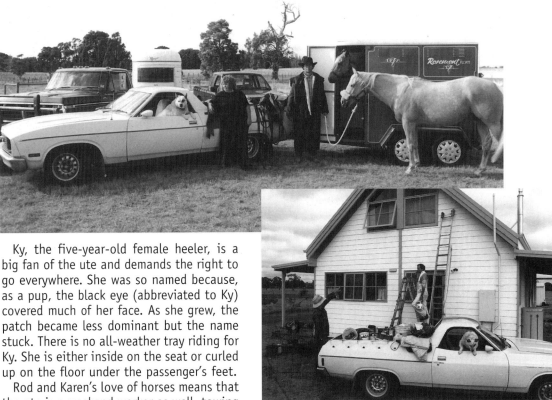

Ky, the five-year-old female heeler, is a big fan of the ute and demands the right to go everywhere. She was so named because, as a pup, the black eye (abbreviated to Ky) covered much of her face. As she grew, the patch became less dominant but the name stuck. There is no all-weather tray riding for Ky. She is either inside on the seat or curled up on the floor under the passenger's feet.

Rod and Karen's love of horses means that the ute is a weekend worker as well, towing a two-horse float. Rod's horse Corop is an Ash Wednesday bushfire orphan and thinks he is human. Karen's horse Dakota is a new addition. Karen is so protective of Dakota that the horse has been nicknamed Precious by members of the riding club of which Karen is president.

Something tells me there will always be a ute in this family for both business and leisure. Ky wouldn't have it any other way!

Vikki Groves has owned two utes. The first was an HT Holden 253 V8 with turbo 400. She bought it when she was about 21 and sold it because it was too expensive to run. She had previously owned a Ford Escort. Her first car was a Mini. Now Vikki has another ute. An 1984 Datsun 1200, with 193,000 kilometres on the clock.

'I bought it out of the Melbourne *Trading Post* less than a couple of years ago, and since then I've spent a lot on it. It's had 'everything' done to it. I've done all the motor mechanicals myself, with the help of my father. You name it, I fixed it — replaced the head gasket, tailshaft and universal. Radiator's been replaced, and hoses and all that. I'm currently changing the interior, and rear tyres. I still have to add front disc brakes to it.

'I could sell it very easily; many people have wanted to buy it. I bought it for $4500, and spent heaps on it.'

Vikki is from the small town of Tallygaroopna, 17 km north of Shepparton in the Goulburn Valley, Victoria. She used to work in Melbourne, travelling home each weekend but now works nearer home.

The Datsun is a great little pale-yellow rig and very economical to run. 'Costs about $25 to and from Melbourne and still

with a few days running around to spare.'

And what does she think of the ute after owning a big V8? 'I love it. Like the look of it. It's cute!'

Vikki is a fully qualified refrigeration mechanic, following in her father's footsteps. He's been doing it for over 30 years.

'I was Dad's shadow. I always watched him work on things, such as his motorbike. I used to sit in the shed and play with his bike, and he'd find the battery flat next morning. He taught me a lot. I've always been interested in motorbikes. The ute has even carried a 750 ZXR Kawasaki in the back, and that was a lot of weight.'

Vikki says her ute has also carried a

lot of gear such as camping gear, and she recently moved her entire house in it over numerous trips.

'I've never really had any problems with owning a ute. No one gave me a hard time. It's been the same with being a female mechanic in a male-dominated career. If anything, I put pressure on myself. I'd hate to have to ask anyone to lift something or do something I couldn't do myself.'

Vikki seems pretty confident and down-to-earth. She is simply doing what she loves and thinks nothing of it. She has a sister who is a singer, and a brother still at high school. She says she never ever wanted to do anything except follow in her father's footsteps. I think he'd be mighty proud.

WANDERING SPARKIE

In 1952, at the age of 15, John Martin of Adelaide began work as a store boy with Mayfields Electrical Contractors. They made a good choice in employing him. He began his electrical apprenticeship the next year and stayed with Mayfields for 45 years and five months, ending his career as electrical supervisor.

His work would take him to many sites such as factories, quarry plants, school extensions, petrol and oil companies, hospitals, multi-storey buildings, power stations, government departments, carparks, town halls, cement factories, police stations, car manufacturers. 'My work varied from high and medium to low voltage in all sorts of work conditions.'

He drove all sorts of utes. 'The best utes I drove were Holdens. The worst was a Falcon back in 1989; it was slow on pick up from starting. If you needed to go fast it wouldn't do it straight away, so you had to allow for the slow start. And it was not until more recent times that the utes were air-conditioned.'

Because John had to work in many remote sites, many of his utes had racks which were loaded with gear such as pipes, ladders, switchboards, cable drums, and all manner of electrical equipment. Depending on the job, he also towed a trailer. Utes were sometimes used to tow heavy cables on some of the larger jobs.

'The utes were used for all sorts of things, I've slept in the back of a ute on a stretcher with the canvas cover over the top. I carried money to pay the workers' wages.'

John would drive to locations all around Adelaide but also made many long-haul trips to places such as Woomera Rocket Range, Roxby Downs, Moomba Gas Fields, Whyalla, Burra, Ceduna, Naracoorte, Coober Pedy, Kangaroo Island, Olympic Dam, Port Lincoln, York Peninsula, Port Augusta, Leigh Creek and Tailem Bend.

He also worked in Victoria at Horsham, in New South Wales at Menindee and Broken Hill, to the west at Kwinana in Western Australia, north at Mt Isa in Queensland, as well as jobs in the Northern Territory at Alice Springs, Tennant Creek, Katherine, Jabiru and Darwin. He was one of those who worked on the rebuilding of Darwin after Cyclone Tracy destroyed the city.

John Martin has seen much of this great country. When he travelled in years past the roads were not the great highways we know of today, but many were dirt roads and some rough tracks. He has memories to last more than a lifetime.

One man spent ten years restoring it, and even then it was not complete. His son took over but he wasn't really interested. Then Wayne Frankcombe bought it from an ad in a paper for $6,000. He had always wanted a 1929 Chev ute, but could never afford one. It took him 15 years before he got what he really wanted and, when he finally did, it was 90 per cent complete, but still in a great heap of bits. It took him a further four years to assemble and restore it as spare money was available. Now it has done about 70,000 kilometres.

Wayne drives it to work as a self-employed registered builder, working on houses, renovations and extensions. He is married to Leanne who drives the family Commodore. 'The ute's not her thing. Too many people stare at her when she's in the ute. People always say 'G'day' at stoplights, or give a wave.'

Son Brent, 18, has a 76 Chev ute, and Luke 17 has a VW Beetle convertible. In the garage being restored is a 34 Chev coupé which Wayne hopes to have on the road by the end of this year. Wayne and some friends are planning to produce 34 coupés in fibreglass and hope to expand this work into a business.

Wayne has long been a ute man. He's had two FC Holden utes, an HR ute, a KB100 Inter ute, a 63 F100 ute, a Mini-Moke ute, a 51 Dodge Fargo ute, and 'can't think of the rest. I guess I just love utes.'

'Working Man's Rolls Royce'

They used to refer to Armstrong Siddeley vehicles by the above name, according to Reg Gobell of Gawler, South Australia. When he was 21 years old in 1949, he would have preferred to have bought a new Chev ute, but there was a three-year waiting list, so he paid £1275 for the new 6-cylinder Armstrong ute, which he only had to wait six weeks to get.

He thought it was pretty plain and not up to the standard of the sedans, but it was brand-new for the young builder, who used it daily for work, loading it up with building gear, a concrete mixer, scaffolding, planks, and tools.

His first ute at 18 had been an old 1927 T-model Ford that he paid £3 for and towed home. He also bought three extra motors. The whole lot cost him £20. He got it working then built a new cabin and sides for the rear tray. He drove it for three years, before saving enough money to buy the Armstrong Siddeley.

As well as using it for work, it was also very much used for pleasure, often carrying many friends to and from dances over the years, with everyone piled into the front cabin, and on the back. He says it was capable of doing 85 mph back then.

He sold it in 1957 trading it in on a Ford V8 Starmodel which he thrashed. Police once pulled him over for doing 100 mph for which he was fined. The Amstrong Siddeley that he had sold passed through a couple of owners before it was bought in 1972 by Reg's father-in-law. In 1976 it was given back to Reg. His father-in-law died in 1977. By the time it had come full circle Reg said it was in reasonable condition, with a few dents in the mudguards.

He set to and reconditioned the motor, added new genuine English leather upholstery, hood lining, and did all the electricals — starter motor, generator, distributor. The original 4-speed synchromesh gearbox was still good.

In 1997, it won the 'Most Original Ute' in the Aussie Ute Muster at Birdwood.

Nowadays it is used once a month or so, and driven at a 'good speed' of 45 mph. A bit different to the old days. His daily vehicle is a red 1988 GLS Ford ute. He's always had utes: 5 Valiants, a 1963 Zephyr, XE and XF Ford Falcons. He worked as a self-employed builder for 50 years before retiring a few years ago. These days he is a member of the Armstrong Siddeley Club, and Gawler Vintage Club, with his son Kym. Eventually Kym will take over the ute when Reg is finished enjoying it.

In the early days when Reg's father worked at the Ammunition Factory in Salisbury, Reg made an unusual piece for the ute. His father got him a butterfly switch from an Avro Anson aeroplane, and fitted it to a brass 25 lb ammunition shell, over the exhaust. It was connected to a lever underneath the seat with a cable leading to the butterfly to open the exhaust. Out on the highway it had a totally illegal and loud, deep throaty tone.

Reg laughs and says 'You could hear it coming down the main road out of town but then I'd flick the switch and it would come into town as quiet as a mouse. Police used to park in private cars about 600 yards from the dance hall to catch us.'

WBs

Driving a WB Holden is like putting up a sign with flashing lights which says, 'Cops, come and get me.'

'Coppers hate WBs, big bullbars, lights, aerials — they just harass you all the time. They think all WB drivers are hoons. They always try to put you off the road.'

I met a bloke in the Western District, a shearer, who told me that he left his farm to go to work six times in six weeks and each time he had been pulled up by the cops. 'They think they'll beat me. Now I just laugh at them; the more they harass me the more determined I am to look after it and keep it on the road.'

His mother later confirmed what he had told me. 'He's a non-drinker, a hard worker, and loves his ute. They are forever giving him a hard time. He was pulled over in New South Wales whilst he was away shearing. His father had to lend him our car and bring his ute home, because they said the ute was unroadworthy. When he took it in they said it was fine and it was straight back on the road.'

Ute lovers seem to have an affinity with WBs. Some may never want to own one, but understand why they have become a cult car all on their own. Many will tell you if

you have a ute, you should have a WB first, that they are the passage to your right to call yourself a 'real' ute owner. Those who first owned a WB have fond memories, even if they have gone on to fancier machines later.

To start off as a WB owner is like earning a badge, an unspoken entrance card to 'the club'. It is a sign that you are really 'fair dinkum'. Many WBs have the guts thrashed out of them or are abused. They are always slicked up. A WB *must* have a bullbar, aerials, big lights and be plastered with stickers. Some people look on them in disgust and think of them as 'wank mobiles', others think of them as a young bloke's holy grail.

The funbusters (police) need to be a bit fairer. Sure there are hoons out there driving WBs, but also a lot of hard-working young people who simply happen to like the look and style of the bush vehicle. 'By trying to put them off the road it is also possible they are putting them out of work — and no one deserves that.'

One bloke pulled a wad of papers from the glovebox. 'See that, they are the collection of roadworthies I've had to have because of the ute. They look for anything. One cop said the lights were illegal. I had to have them checked and lose a day's pay running around, only to be told he had no right, and that there was nothing wrong with them. Bastards. I'll fight them forever.'

Wherever I went it was the same. Young people often didn't have a very good attitude towards police because of police harassment. Now 'we work in packs and look after each other. After all, we have earned the right to drive.'

Paul and Clinton Turnbull both got their first utes when they turned 17.

Twenty-one-year-old Paul has had three utes in three years. His first was a WB Holden, then a VP Commodore, and now the VS Commodore. His VP stayed in the family because he sold it to 17-year-old little brother, Clinton.

Paul is in his fourth and final year at Dookie Agricultural College studying for a degree in Agricultural Science. He hopes to get a job in Agronomy, such as a field officer, but eventually wants to move back to work on the farm.

'There's such a jump up from the WB to VP then VS, both for comfort and performance,' he says. 'The WB which I bought privately from an advert in the paper was a neat street ute. But I was spending too much money on it so decided to get something more reliable.'

The 1992 Pascal Blue VP Holden Commodore with V6 motor, bullbar, lights, UHF radio and Club Sport wheels followed. He sold it to Clinton last year. Clinton says he bought it because, 'I know what's been done to it, and it was a good deal. Better than I expected.'

Paul moved up into the Botanica Green 1996 VS 'Maloo' Commodore when Clinton took the VP. It was Clinton who saw the VS

first in a caryard in Geelong. Paul took it for a test drive, liked it and two weeks later it was his. He's pretty impressed with it, and will stick to Holdens. Both he and Clinton agree that Holdens 'look better' and have good resale value.

I asked their mother Pam, at the kitchen sink, what she thought of all these utes. She laughed, but it was the shy smiles on the two boys' faces that told me — Mum was a real ute lover too.

'When I take a ute to town, there's always comments like "oh, you've got the ute today",' she laughs. 'I like them because they are good to drive. I like the power.'

The two boys laugh when I jokingly ask what's it like to have a Mum who's a 'hoon'. Paul says, 'if we get into trouble she can't say much'. I didn't get to ask Dad what he thought because he had left to drive his father into town, but it was obvious he was a ute man too from way back. You just can't imagine how important a ute is for people such as the Turnbulls.

Clinton is in year 12, his final year at school, before he too hopes to go to Dookie to study for a farming degree, and then return to the farm. Clinton didn't say it but I could tell he was counting the days until his 18th birthday when he'll get his licence, then he and the VP can hit the road.

Acknowledgements...

Bob Jane Corporation (Mark Brown, Greg Wheeler), Duhig Ford Essendon (Ian, Eddy, Craig), Ford Australia Public Affairs (Pam), Kevin Grinham, Guildford Store (John and Jacinta Houlihan), Rob Hickman, Tony & Lisa Marinelli, National Motor Museum (Jon Chittleborough, Brenton Moar), Dot and Colin Nesbit, Nu-Color-Vue (Margaret), Smith's Kodak Express (Lee & Tasha), Karen Stedman, and also many thanks to Graham Williams (who kept my computers mobile).

My sincere thanks to each and every one of the people in this book and I thank them for their generosity. To those people whose stories have had to be deleted due to space constraints I can only say I'm sorry and hope they may appear in a future book.

Special thanks to designer Geoff Hocking for this, the sixth book we've worked together on; to Kevin Masman for his computer skills; and to Karen Masman for first draft editorial work. Their commitment is greatly appreciated.

Some photo's from private collections were copied on site. I wish to acknowledge the co-operation of the owners of the following photos: pp. 38, 39, 61, 69 (top), 74, 90 (lower), 91, 93 (top), 98, 101, 117, 120, 125 (top & bottom left), 142 (right), 145, 146, 147, 154.

All other photographs were taken by the author using a secondhand 1975 Canon FT 35 mm with wide angle lens, Kodak Gold 100 film stock, and using natural light only.

Allan M. Nixon travelled over 32,000 kilometres looking at utes. He lives with his wife, Janette, at the base of a mountain range north of Melbourne, with two dogs Rowdy and Lady, a possum, two chooks and many native birds. He likes to 'go bush' and spends a lot of time 'on the road'.

He is currently researching two new books, as well as collecting utes for a *Beaut UTES* Museum.

Books by Allan M. Nixon

Inglewood: Gold Town of Early Victoria.
Muddy Boots: Inglewood Football Club.
Inglewood Gold: 1859–1982.
100 Australian Bushrangers: 1789–1901.
The Grinham Report: A Family History.
The Swagmen: Survivors of the Great Depression.
Somewhere In France: Sgt. Roy Whitelaw. A.I.F. 1914–1918.
Stand & Deliver: 100 Australian Bushrangers.
Pocket Positives: An A–Z of Inspirational Quotations.
Humping Bluey: Swagmen of Australia.